Compendium of Stone Fruit Diseases

Edited by

Joseph M. Ogawa
University of California
Davis

Eldon I. Zehr
Clemson University
Clemson, South Carolina

George W. Bird
Michigan State University
East Lansing

David F. Ritchie
North Carolina State University
Raleigh

Kiyoto Uriu
University of California
Davis

Jerry K. Uyemoto
University of California
Davis

APS PRESS
The American Phytopathological Society

Financial Sponsors

California Cling Peach Advisory Board
California Prune Board
California Tree Fruit Agreement
Cortez Growers Association, Inc.
Elf Atochem North America, Inc.
Fowler Nurseries, Inc.
Giannini Packing Corporation
Gowan Company
Griffin Corporation
Ito Packing Company, Inc.
John Kashiki, Chairman, KASH, Inc.
Monsanto Agricultural Group
NOR-AM Chemical Company
O-G Packing
Rhône-Poulenc Ag Company
Rohm and Haas Company
Sandoz Agro, Inc.
Tufts Ranch
Zeneca Ag Products

Cover photographs courtesy J. M. Ogawa
Front: peaches and apricot blossoms (inset); back (clockwise, from upper left): apricots, prune blossom, nectarine, and sweet cherry blossom

Reference in this publication to a trademark, proprietary product, or company name by personnel of the U.S. Department of Agriculture or anyone else is intended for explicit description only and does not imply approval or recommendation to the exclusion of others that may be suitable.

Library of Congress Catalog Card Number: 95-77906
International Standard Book Number: 0-89054-174-4

Printed in the United States of America

The American Phytopathological Society
3340 Pilot Knob Road
St. Paul, MN 55121-2097, USA

Preface

A *compendium* is a brief treatment or summary of a larger body of knowledge. As with an abstract, a compendium is difficult to prepare—to convey fairly and accurately a large amount of accumulated knowledge without omitting facts and details that readers need or wish to know. The editors of this compendium do not have expertise in all of the diseases and disorders described herein. As has been done in other compendia, we relied on others, in the United States and elsewhere, to write most of the information assembled here. We are indebted to these scientists who responded willingly and with considerable patience to our requests. Those who prepared the descriptions are listed at the end of each section. Contributors of photographs and illustrations are acknowledged only when they were not the authors of the accompanying text.

Our task as editors was to solicit contributions, assemble the material, ascertain that each topic was addressed in a coherent, concise manner, and verify that the information was current and complete. In this task we were assisted by others who served as reviewers for each manuscript that was submitted. In addition to our own critical review, each description was reviewed whenever possible by two persons known to have first-hand knowledge of the disease or disorder described.

We gratefully acknowledge the following individuals, who prepared sections of the compendium. Those who reviewed portions of the manuscript as well are identified with an asterisk.

A. N. Adams, AFRC Institute of Horticultural Research, East Malling, Maidstone, Kent, England
J. E. Adaskaveg,* Department of Plant Pathology, University of California, Davis
M. D. Aichele, Yakima, Washington
T. A. Baugher, West Virginia University Experiment Farm, Kearneysville
B. Bernstein, Department of Plant Pathology, Clemson University, Clemson, South Carolina
P. F. Bertrand,* Cooperative Extension, University of Georgia, Tifton
A. R. Biggs,* University Experiment Farm, West Virginia University, Kearneysville
G. W. Bird, Department of Entomology, Michigan State University, East Lansing
G. T. Browne, Agricultural Extension Service, University of California, Bakersfield
T. J. Burr, Department of Plant Pathology, New York State Agricultural Experiment Station, Cornell University, Geneva
M. V. Carter, Hawthorndene, South Australia
J. E. DeVay,* Department of Plant Pathology, University of California, Davis
H. English,* Department of Plant Pathology, University of California, Davis
A. J. Feliciano, Department of Plant Pathology, University of California, Davis
K. M. Flint, Department of Biology, Utah State University, Logan
J. A. Foster, USDA-APHIS-PPQ, Beltsville, Maryland
R. L. Gilbertson,* Department of Plant Pathology, University of Arizona, Tucson
D. Gonsalves, Department of Plant Pathology, New York State Agricultural Experiment Station, Cornell University, Geneva
G. G. Grove,* Tree Fruit Research and Extension Center, Washington State University, Wenatchee
W. D. Gubler,* Department of Plant Pathology, University of California, Davis
A. J. Hansen, Agriculture Canada, Research Station, Summerland, British Columbia, Canada
Y. Harada,* Faculty of Agriculture, Hirosaki University, Hirosaki, Aomori, Japan
M. J. Hattingh, University of Stellenbosch, Stellenbosch, South Africa
F. F. Hendrix, Jr.,* Department of Plant Pathology, University of Georgia, Athens
K. D. Hickey,* Fruit Research Laboratory, Pennsylvania State University, Biglerville
A. F. Iezzoni, Department of Horticulture, Michigan State University, East Lansing
A. L. Jones,* Department of Botany and Plant Pathology, Michigan State University, East Lansing
B. C. Kirkpatrick, Department of Plant Pathology, University of California, Davis
H. Kitajima,* Institute for Plant Virus Research, Tsukuba-gun, Ibaraki, Japan
H. J. Larsen, Orchard Mesa Research Center, Colorado State University, Grand Junction
S. D. Lyda, Department of Plant Pathology, Texas A&M University, College Station
H. Melakeberhan, Department of Entomology, Michigan State University, East Lansing
R. W. Miller,* Department of Plant Pathology and Physiology, Clemson University, Clemson, South Carolina
G. I. Mink,* Department of Plant Pathology, Washington State University, Prosser
S. M. Mircetich, USDA-ARS, Department of Plant Pathology, University of California, Davis
J. M. Ogawa,* Department of Plant Pathology, University of California, Davis
G. N. Oldfield, Department of Plant Pathology, University of California, Riverside
C. L. Parish, U.S. Department of Agriculture, Tree Fruit Research and Extension Center, Wenatchee, Washington
J. W. Pscheidt,* Department of Botany and Plant Pathology, Oregon State University, Corvallis
A. H. Purcell, Department of Entomology, University of California, Berkeley
P. L. Pusey,* USDA-ARS, Tree Fruit Research and Extension Center, Wenatchee, Washington
D. C. Ramsdell, Department of Botany and Plant Pathology, Michigan State University, East Lansing
P. R. Rener, Department of Plant Pathology, University of California, Davis

D. C. Ramsdell, Department of Botany and Plant Pathology, Michigan State University, East Lansing
P. R. Rener, Department of Plant Pathology, University of California, Davis
J. D. Ridley, Department of Horticulture, Clemson University, Clemson, South Carolina
D. F. Ritchie,* Department of Plant Pathology, North Carolina State University, Raleigh
I. M. M. Roos, Fruit and Fruit Technology Research Institute, Stellenbosch, South Africa
K. Ryugo, Department of Pomology, University of California, Davis
K. Sawamura, Faculty of Agriculture, Hirosaki University, Hirosaki, Aomori-ken, Japan
E. Seemüller, Federal Biological Research Centre for Agriculture and Forestry, Institute for Plant Protection in Fruit Crops, Dossenheim, Germany
S. M. Southwick,* Department of Pomology, University of California, Davis
A. Sztejnberg, Department of Plant Pathology and Microbiology, Faculty of Agriculture, Rehovot Hebrew University of Jerusalem, Rehovot, Israel
K. G. Tate, MAF Technology North Central, Hawke's Bay Agricultural Research Centre, Hastings, New Zealand
Y. Terai,* Yamanashi Fruit Tree Experiment Station, Yamanashi, Japan
K. Uriu,* Department of Pomology, University of California, Davis
J. K. Uyemoto,* USDA-ARS, Department of Plant Pathology, University of California, Davis
H. E. Waterworth, USDA, National Plant Germplasm Quarantine Laboratory, Glenn Dale, Maryland
J. M. Wells, USDA-ARS, Philadelphia, Pennsylvania
Y. Wu, Horticultural Research Institute, Shanghai Academy of Agricultural Sciences, Shanghai, China
J. M. Young, Department of Scientific and Industrial Research, Auckland, New Zealand
E. I. Zehr,* Department of Plant Pathology and Physiology, Clemson University, Clemson, South Carolina

Those who served solely as reviewers are as follows:

J. Amador, Texas Agricultural Extension Service, Weslaco
J. L. Anderson, Department of Plants, Soils, and Biometeorology, Utah State University, Logan
R. L. Anderson, Department of Pomology, Cornell University, Ithaca, New York
A. N. Azarenko, Department of Horticulture, Oregon State University, Corvallis
E. E. Butler, Department of Plant Pathology, University of California, Davis
R. M. Carlson, Department of Pomology, University of California, Davis
D. R. Cooley, Department of Plant Pathology, University of Massachusetts, Amherst
J. A. Hopfinger, Rutgers Cooperative Extension, Cook College, New Brunswick, New Jersey
D. L. Hopkins, Central Florida Research and Education Center, University of Florida, Leesburg
B. A. Jaffee, Department of Nematology, University of California, Davis
S. N. Jeffers, Ecoscience Corp., Worcester, Massachusetts
K. B. Johnson, Department of Botany and Plant Pathology, Oregon State University, Corvallis
R. S. Johnson, Department of Pomology, University of California, Davis
A. J. Latham, Department of Plant Pathology, Auburn University, Auburn, Alabama
B. T. Manji, Department of Plant Pathology, University of California, Davis
M. V. McKenry, Kearney Agricultural Center, University of California, Parlier
I. Merwin, Department of Fruit and Vegetable Science, Cornell University, Ithaca, New York
S. S. Miller, USDA-ARS, Appalachian Fruit Research Station, Kearneysville, West Virginia
A. P. Nyczepir, USDA-ARS, Southeastern Fruit and Tree Nut Research Laboratory, Byron, Georgia
G. L. Philley, Texas Agricultural Extension Service, Overton
D. J. Phillips, USDA-ARS, Fresno, California
E. L. Proebsting, Irrigated Agriculture Research and Extension Center, Washington State University, Prosser
D. Ramos, Department of Pomology, University of California, Davis
S. M. Ries, University of Illinois, Urbana
W. C. Schnathorst, Department of Plant Pathology, University of California, Davis
S. Scott, Department of Plant Pathology and Physiology, Clemson University, Clemson, South Carolina
R. M. Sonoda, Agricultural Research and Education Center, University of Florida, Fort Pierce
J. K. Springer, Rutgers Research and Development Center, Bridgeton, New Jersey
P. W. Steiner, Department of Botany, University of Maryland, College Park
R. Utkhede, Entomology/Plant Pathology, Research Station, Summerland, British Columbia, Canada
W. F. Wilcox, Department of Plant Pathology, New York State Agricultural Experiment Station, Cornell University, Geneva
K. S. Yoder, Winchester Agricultural Experiment Station, Virginia Polytechnic Institute and State University, Winchester

We thank the departments of each of the coeditors for the role they played in providing services required for the development of this compendium, including stenographic, photographic, and library services. Pamela R. Rener, laboratory assistant on fruit crops in the Department of Plant Pathology at the University of California, Davis, acted as right-hand assistant to the editors of this compendium in ensuring the orderly compilation of manuscripts.

Special acknowledgment is extended to the staff at APS headquarters; with specific thanks to Richard W. Smiley and Steven A. Slack, former APS Press Editors-in-Chief; and to Joyce E. Loper, member of the APS Press Editorial Board assigned to this publication.

Contents

Part II. Noninfectious Disorders

Introduction

Apricot

Origin and Production

The apricot, *Prunus armeniaca* L., was first cultivated about 4,000 years ago in China. As a result of commerce, apricots were introduced into southwest Asia (Armenia), then into Italy by 100 B.C., into England by 1620 A.D., and into the United States (state of Virginia) about 1629. The former USSR produces the most apricots annually, followed by Turkey, Italy, Spain, Greece, France, and the United States. California produces 97% (9,308 ha) of the U.S. crop, primarily in the San Joaquin Valley. Other states with apricot production include Washington, Utah, Michigan, and Arizona.

Botany and Biology

Most cultivated apricots belong to the species *Prunus armeniaca* L., which is diploid ($2n = 16$). Flowers are borne singly or doubly at a node, with five sepals and petals, about 30 stamens, and one pistil. The fruit is a drupe consisting of a stony endocarp, a fleshy mesocarp, and an exocarp. Ripe fruit are soft to touch and highly susceptible to decay-causing organisms. They are picked at different maturity stages depending upon market outlet, cultivar, and growing district. Fruit can be stored for up to three to four weeks near 0°C and 85% relative humidity. Most commercial cultivars are self-fertile, but several are self-incompatible. Floral buds are initiated in late spring or summer. The chilling required to initiate flowering (below 7.2°C) ranges from 300 to 1,200 hr; in years with insufficient winter chilling during December and January, bud drop can occur. The bloom period is one to two weeks in early spring. On several cultivars, fruit thinning is required to obtain adequate fruit size. Dormant trees tolerate winter cold, but early emergence from dormancy results in freeze injury to blossoms and death of trees in many growing areas where late freezes occur. Consequently, the production area is limited, because early bloom makes apricots prone to frost injury.

Horticulture

Cultivars

Cultivars widely planted around the world include Hacihaliloglu, Tokaloglu, Bulida, Canino, Moniqui, Polonais, Bergeron, Rouge du Rousillon, Hungarian Best, Reale d'Imola, Monaco Bello, San Castrese, Tyrinthos, Bebeco, Luizet, Cafona, Palumella, Sundrop, Moorpark, Trevatt, Palsteyn, Umberto, Yerevan, and Maj-Ho-sin. The predominant cultivar grown commercially in California (cv. Blenheim, syn. Royal) originated in the United Kingdom and France. Other cultivars grown in California include Tilton and Patterson.

Rootstocks

Apricots usually are propagated by budding a commercial cultivar onto apricot seedlings, Lovell or Nemaguard peach seedlings, Myrobalan plum seedlings, or rooted cuttings of plum cultivar Marianna 2624. No dwarfing rootstocks are used commercially. Peach and apricot rootstocks are highly susceptible to Phytophthora root rot; Myrobalan plum should be used in heavy, poorly drained soils, and Marianna 2624 in heavy soil or sites infested with Armillaria root rot organism.

Culture

Common tree spacing of 7.3 m in deep, fertile soil favors use of equipment for spraying trees to control pests and diseases. Fruit production begins in the third year and reaches maximum by the sixth or seventh year but may continue even after 60 years. Pruning is necessary to maintain form, fruit production, and tree vigor.

Harvest and Handling

All fruit for fresh market are hand-harvested. Mechanical trunk shakers and catching frames used for processed fruit may increase trunk injuries and incidence of canker diseases. In California, the crop is utilized as canned (50%), dried (22%), frozen (10%), fresh market (15%), and baby food (3%). Other products are wine, brandy, jam, and nectar. Ground apricot pits are used to clean jet engines, and the kernel oil is used for soaps and perfume. When fruits are to be dried, mature fruit are held for further ripening, treated with sulfur dioxide, and placed on wooden trays in the sun.

Selected References

Ogawa, J. M., and English, H. 1991. Diseases of Temperate Zone Tree Fruit and Nut Crops. Publ. 3345. University of California, Division of Agriculture and Natural Resources, Oakland, CA.

Ryugo, K. 1988. Fruit Culture: Its Science and Art. John Wiley & Sons, New York.

Stokes, G., Southwick, S. M., Sun Garden Packing Company, Mariani Packing Company, J. M. Smucker Company, and Zaiger's Genetics. 1992. Apricots and apricot processing. Pages 93-103 in: Encyclopedia of Food Service and Technology, Vol. I. Y. H. Hui, ed. John Wiley & Sons, New York.

(Prepared by J. M. Ogawa and S. M. Southwick)

Peach and Nectarine

Origin and Production

The peach, *Prunus persica* (L.) Batsch, originated in China, where records of cultivation date back 3,000 years. Peaches probably were brought to Persia through mountain trading routes, becoming known there as the Persian fruit, hence the name *persica*, or peach. These derived terms led to the misconception that peaches originated in Persia.

By 330 B.C., peaches reached Greece, and during the Middle Ages, their culture spread throughout Europe. The Portuguese apparently introduced the peach to the east coast of South America. The Spanish brought peaches to Florida and Mexico, the French to Louisiana, and the English settlers to Virginia and Massachusetts. Yellow-fleshed cling fruit were common in Spain, whereas white-fleshed, soft-textured fruit predominated in France and England. Eventually, the two sources of germ plasm were mixed. Many recent cultivars grown in the United States can be traced to a single cultivar called Chinese Cling (Shanghai Shumi) that was introduced from China in 1850 by Charles Downing and grown in Marshalville, Georgia. The white-fleshed peach Belle of Georgia and the yellow-fleshed Elberta were selected from seedlings that originated from Chinese Cling.

The nectarine, *Prunus persica* (L.) Batsch var. *nucipersica* (Suckow) C.K. Schneid., is a hairless or fuzzless peach, derived by mutation of the dominant gene for fuzziness to the recessive gene for smooth skin. The place of origin of the nectarine is unknown, but its history goes back approximately 2,000 years. The first recorded case of a peach tree bearing a nectarine fruit was in 1741.

Today, peaches and nectarines are important in the agricultural economy of many countries. In 1992, world peach and nectarine production was approximately 10 million metric tons, with Italy accounting for about 19% and the United States for 14% of the total production.

Botany and Biology

Peach and nectarine fruits are produced on one-year-old wood from flower buds formed the previous growing season. Typically, a leaf bud flanked by two flower buds forms in each node. Bud primordia give rise to the next season's leaves, and shoots and flowers are produced in the axils of primary and secondary leaves in the spring and early summer. Flower buds are initiated in summer after growth of new shoots slows. Most commercial cultivars have upright to moderately spreading growth habit. Cultivars vary in cold-hardiness and require from less than 100 to more than 1,000 chilling hours (accumulated hours below 7.2°C) for normal bud development.

Nearly all commercial cultivars of peaches and nectarines are self-fertile. Fertilization usually occurs 24–48 hr after pollination. The fruit is a drupe with a thin epidermis, a fleshy mesocarp, and a stony endocarp containing the seed. The fleshy mesocarp may adhere to the stone (clingstone) or separate from it (freestone), and the color may be white, yellow, orange, or red. The mesocarp of ripening fruit may soften (melting), as is typical of fresh market fruit, or remain firm (nonmelting), as is common for canning.

Horticulture

Cultivars

Few fruit species are grown in as many diverse climatic conditions as peaches and nectarines. Although considered temperate fruits, their commercial production is being explored in the tropics and subtropical regions. The trees' climatic requirements, the fruit's short shelf life, which necessitates cultivars ripening at different seasons, consumers' preferences for flesh color, and the fruit's intended use (fresh market, canning, freezing, or drying) contribute to the diversity and large number of cultivars grown throughout the world.

Some important peach cultivars with melting flesh are Redhaven, Elberta, O'Henry, Elegant Lady, and Springcrest; some with nonmelting flesh are Andross, Kanto-5, and Fonghuag. Examples of nectarine cultivars are Fantasia and Flavortop.

Rootstocks

Rootstocks from peach seedlings, interspecific peach hybrids, plums, and plum hybrids are used worldwide for peaches and nectarines. Peach seedlings, the most widely used, may be obtained from wild selections. Tennessee Naturals or Indian Peach (United States), Franco (Spain and Italy), and Creole (Mexico) are the dominant sources of peach seedling rootstocks. Commercial cultivars utilized in the processing or drying industry, like Halford and Lovell (United States, Canada, and Mexico), Kakamas (South Africa), and Capdeboscq (Brazil), are important also. Siberian C was selected for cold-hardiness and is being used in regions with severe winters (e.g., Canada). Okinawa is being utilized in Australia, South Africa, Japan, and in areas with very mild winters where root-knot nematodes are a problem. Interspecific hybrids of peach and almond (e.g., GF 556 and GF 677 from France) have excellent adaptability to alkaline, wet, or dry soils and are vigorous in replant soils. Nemaguard and Nemared, which are *P. persica* × *P. davidiana* hybrids, are used extensively in California and Florida and in some other areas where root-knot nematodes are a problem. Plums and plum hybrids are better adapted to wet, waterlogged soils and are being used where peach and interspecific peach hybrids are not suitable. The disadvantages of plum rootstocks are graft incompatibility, short tree life, suckering, and low fertility.

Culture

Like apricot, peach and nectarine are sensitive to low temperatures. The trees perform best on sites that have fertile, deep, and naturally well-drained soils and good ventilation. In almost all soils in humid areas, liming (to raise soil pH to about 6.5) and fertilizer additions are required for optimum production. Nitrogen, the key nutrient that controls growth and fruiting, is applied annually. Irrigation is used widely in semiarid production areas.

Peaches and nectarines are traditionally trained to an open-center or vase-shaped system. Recently, more intensive planting systems to maximize the use of land, promote fruiting precocity, and improve utility for mechanization are gaining popularity. Because dwarfing rootstocks are not used, intensive planting is possible only through appropriate tree training and support. With the intensive Tatura trellis system, as many as 1,500–2,500 trees per hectare can be accommodated, compared with 250–450 trees with the open-center system. Two popular high-density planting systems in California are the cordon and the free-standing perpendicular V (900–1,200 trees).

Peaches and nectarines require annual pruning; the type and intensity of pruning depend on the planting system used. Trees usually set too many fruit and require hand thinning to obtain fruits of marketable size.

Harvest and Handling

The maturity indexes commonly used for peaches and nectarines are color and firmness. The best time to harvest for shipping is probably just as the ground color begins to change to yellow for yellow-flesh cultivars and to white for white-flesh cultivars. Color, however, fluctuates from year to year and is best used in conjunction with firmness, which is measured in the United States in pounds, with a 5/16-in. pressure tester. Optimum firmness at harvest for high-quality fruit in the marketplace varies with cultivar.

Peaches and nectarines are highly perishable and require careful handling at harvest and after. Fruits are picked almost entirely by hand, although in the United States, about 25% of processing peaches are harvested mechanically. In the tropics and subtropics, ripening continues over two or three weeks, and each cultivar is picked five or more times. Ripening is more uniform in temperate regions, but an orchard still is harvested two or three times. In the United States, fruit for fresh market commonly are transported to a central packing facility, where rapid cooling, rinsing in potable water, defuzzing, waxing, and packing are done. The defuzzing and wax-oil treatment are to improve appearance and reduce moisture loss during shipment. Fungicides may be added to the wax-oil treatment to control fruit decay.

Selected References

Childers, N. F., and Sherman, W. B., eds. 1988. The Peach: Culture, Cultivars, Breeding, Propagation, Nutrition, Training and Pruning, Diseases and Insects, Harvesting, Storage and Marketing. 4th ed. Horticultural Publications, Gainesville, FL. 986 pp.

Rom, R. C., and Carlson, F. 1987. Rootstocks for Fruit Crops. Wiley, New York. 494 pp.

(Prepared by A. J. Feliciano)

Plum and Prune

Origin and Production

Although some plum species are native to North America, commercial cultivars grown in the United States originated in Europe and Japan. Plums are classified into two groups: the European plum, *Prunus domestica* L., was introduced from Europe; the Japanese plum, *Prunus salicina* Lindl., is native to China and was introduced from Japan. The leading plum-producing regions (followed by annual yields in 1,000 t) are the former Soviet Union (1,000), Romania (800), China (765), the former Yugoslavia (765), and the United States (680). Other important plum and prune production countries are Germany, France, Turkey, Hungary, Italy, Spain, Bulgaria, Chile, South Africa, and Australia. The 17,000 ha grown in California accounts for over 90% of the production in the United States.

A prune, by definition, is a dried plum. Certain European or "domestica" plums can be dried whole without fermenting because of their high sugar content. The French prune constitutes approximately 96% (40,000 ha) of California's prune production. California's production constitutes 70% of the world's prune supply and 99% of that of the United States.

Botany and Biology

Pollination requirements of plums vary from completely self-incompatible to completely self-compatible. Many commercial Japanese cultivars are self-incompatible. Those that are self-incompatible require a suitable pollinizer. Exceptions are the partially or wholly self-fruitful cultivars Casselman, Grandrosa, Santa Rosa, Simka, and Roysum. In the temperate climate of California, they bloom in late February or early March and ripen from early May through September, whereas in the northern states, they bloom about a month later and ripen from August through September. The French or La Petite d'Agen is self-fruitful, but the pollen is heavy and sticky, and often stigmas are higher than anthers, making pollen transfer by pollinating insects necessary to ensure consistent cropping.

The flowers in plums appear before the leaves. Flowers grow in clusters of one to three along the new growth in Japanese plums, whereas in European plums (prunes) they are formed in buds that usually contain one or two flowers and no leaves. The buds are formed laterally in the axils of leaves on either the new growth of spurs or, in some cases, the older portions of the current season's growth. The crop is produced mostly on spurs. A fully grown tree may have over 100,000 flowers, but only about 1% fruit set is necessary for a full crop. Removal of fruit by hand is required to obtain fruit of suitable size for marketing.

European and Japanese plums probably originated from the doubling of chromosomes between *P. cerasifera* ($2n$ = 16) and *P. spinosa* ($2n$ = 32), because they possess 48 chromosomes. They can tolerate cold temperatures during dormancy and also hot, arid climates, if irrigated. However, prunes are susceptible to sunburn ("blue prunes") if cool summer temperatures are followed immediately by temperatures exceeding 38°C. Preharvest fruit drop is a concern for many growers.

Horticulture

Cultivars

Plums grown for fresh consumption are mainly cultivars of the Japanese plum and were introduced into California around 1870. The genetic improvement of plums is credited to Luther Burbank, the noted horticulturist who hybridized them in Santa Rosa, California. Today, with over 150 cultivars, the important ones include the old cultivar Santa Rosa plus new introductions such as Casselman, Laroda, Black Amber, El Dorado, Black Beaut, Red Beaut, Angelino, Nubiana, Simka, and Friar. Leading cultivars in other regions of the United States include Shiro, Methley, Burbank, Ozark Premier, and Burbank Elephant Heart. Other cultivars grown in other parts of the world and in France include French prune (clonal 707), Green Gage, and Reine-Claude. Prunes grown for drying have higher sugar content and are dried in forced-air tunnels heated by gas. The fresh and dry weights (drying ratios) range from 2:1 to 4:1. The most popular cultivar is the California Improved French. Other cultivars include Early Gerrans, Early French, Imperial, Burton, Moyer, Italian, and Friedman. Prunes sold for dessert fruit are Moyer, Stanley, Green Gage, Tragedy, Empress, and President. The fresh market for French prunes has increased, especially in the Asian (Hong Kong, Taiwan, Singapore) market.

Rootstocks

Plums and prunes are propagated by budding onto seedlings or rooted cuttings of Myrobalan plum or peach, or Myrobalan 29C, Marianna 2624, or Nemaguard peach. The Marianna 2624 often produces extensive root suckers. Peach roots can lead to excessive cropping, resulting in smaller fruit or the need for fruit removal. Other popular rootstocks in Europe include INRA Marianna GF 801, INRA peach almond hybrid EF-67, Citation, Pixy, and St. Julien.

Culture

Plums for fresh market can be grown in many soils, but deep, fertile soil is best. Prunes for drying can tolerate poorly drained, fine-textured, marginal soils and tend to be grown there. Different cultivars of Japanese plum are planted in orchard blocks to improve pollination, whereas European French prunes are planted in solid blocks. Production begins in the third or fourth year, increasing to the 12th year, and continues for about 30 years. Trees are generally spaced 18–20 ft between rows. Irrigation procedures have changed from flood and furrow to microsprinkler, drip, and buried drip, and this has prevented excessive irrigation, which favors Phytophthora root rot. Overcropping and sunburning of Japanese and European plums can be regulated to a degree by pruning and fruit removal.

Harvest and Handling

Japanese and European plums harvested for fresh market are picked by hand one to three times, placed in bins for hauling to packing sheds, treated with a wax containing mold inhibitors or left untreated, and packaged in trays or boxes. Plums for drying are harvested with mechanical shakers onto fabric catch-frames; conveyor belts transfer them to positions where forced air can separate the trash from the fruit. The clean fruit drop into bins, which are moved to dehydrators.

Fruit for drying are placed on wooden trays and subjected to a constant flow of warm, circulating air (76.6°C) for 14–24 hr, which reduces the moisture content to 16–20%. After sorting, the moisture content of prunes is restored to 26–32%.

Selected References

Bernhard, R. ed. 1989. Int. Symp. 4th Plum and Prune Genetics, Breeding and Pomology. Acta Hortic. 283, December 1990. 333 pp.

Konishi, K., Iwahori, S., Kitagawa, H., and Yukawa, T. 1994. Horticulture in Japan. Asakura Publishing Co., Ltd. Tokyo. 180 pp.

LaRue, J. H., and Johnson, R. S., eds. 1989. Peaches, Plums and Nectarines—Growing and Handling for Fresh Market. Univ. Calif. Div. Agric. Nat. Res. Pub. 3331. 246 pp.

Ramos, D., ed. 1981. Prune Orchard Management. Univ. Calif. Div. Agric. Sci. Spec. Publ. 3269. 156 pp.

(Prepared by S. M. Southwick and J. M. Ogawa)

Sweet Cherry

Origin and Production

Cherries were first described in 300 B.C. by Theophrastus and had probably been grown several centuries earlier for their wood. The sweet cherry, *Prunus avium* L., is believed to be native to southwestern Asia in the area around the Caspian and Black seas. Beginning in the 16th century, cherries were widely planted in Europe, particularly in Germany. Europeans first brought sweet cherries to North America. The leading regions in production are the former Soviet Union (32.1%), the United States (15.1%), Turkey (9.6%), Germany (8.9%), the former Yugoslavia (4.9%), Hungary, Poland, Romania, and Italy. Other countries with significant produc-tion are France, Spain, Greece, Canada, Japan, Chile, New Zealand, and Australia. Ninety percent of the U.S. crop is grown in California, Oregon, Washington, and Michigan.

Botany and Biology

Sweet cherry belongs to the Rosaceae family and is diploid ($2n$ = 16). Flower buds are formed both on short spurs and at the base of shoots. Fruiting spurs can be productive for up to 10–12 years. Trees occasionally live 100 years. Most sweet cherry cultivars are self-sterile, and some cultivars show intersterility as well. Recently, self-fertility has been introduced into cultivars such as Stella and Lapins. Ordinarily, about 25–30% of the flowers must set fruit in order to produce a full crop. Normally, pollinators are planted every third tree in every third row, and more than one cultivar of pollinator is planted to ensure overlap of bloom. Cherries are pollinated by bees, which are often brought into orchards during bloom in order to increase fruit set.

Horticulture

Cultivars

There are over 200 sweet cherry cultivars worldwide. The most commonly grown cultivars are Bing, Black Tartarian, Brooks, Corum, Early Burlat, Garnet, Jubilee, King, Lambert, Larian, Rainier, Royal Ann (syn. Napoleon), Ruby, Schmidt, Tulare, Van, and Windsor. Royal Ann (syn. Napoleon) and Black Tartarian originated in Europe; Van, Lapins, and Stella were developed in Canada; Rainier was developed in Washington; Bing, Lambert, and Corum originated in Oregon; and Larian and Jubilee were developed in California.

Rootstocks

Only a few rootstocks are available: mahaleb, mazzard, Stockton Morello, and Colt. None is suited to every situation. Mahaleb (*Prunus mahaleb*), known in Europe as St. Lucie, is native to central and southern Europe and is more closely related to *P. cerasus* (sour cherry) than to *P. avium*. Mahaleb is hardier than mazzard, tolerates alkaline soils, and when cropped heavily has a moderate dwarfing effect that often decreases life span. Mazzard (*P. avium*) is the wild sweet cherry from which all cultivated sweet cherries are derived. It forms a large, productive tree but is slow to start bearing. Stockton Morello is a clone of *P. cerasus* suckers, has a moderate dwarfing effect, forms trees that are more spreading than upright, and induces precocity. Colt, a cross between *P. avium* and *P. pseudocerasus*, has substantial resistance to some *Phytophthora* species and apparent resistance to cherry stem pitting virus; however, it is less drought-tolerant than mahaleb. Rootstock breeding in Germany (Giessen series) and Belgium (Gembloux series) introduced rootstocks capable of dwarfing sweet cherry trees and greatly increasing precocious fruiting.

Culture

Site selection is critical for successful cherry production. In most cherries, the chilling requirement varies from 700 to 1,400 hr below 7.2°C. Sweet cherries are susceptible to cold damage. Dormant flower buds are killed by temperatures between −22 and −35°C, depending on the preceding temperature. Sweet cherry flowers are damaged by temperatures of −2°C. Late winter freezes and early spring frosts are most frequently the limiting factors in cherry production. Cultivation in areas with heavy rainfall during fruit harvest should be avoided due to the propensity for fruit cracking. Most commercial cultivars grown in the western United States are highly susceptible to rain cracking, whereas Napoleon, Stella, and Rainier are intermediate in susceptibility, and Black Tartarian, Sam, and Jubilee are among the least susceptible cultivars.

Areas with extremely high average daily summer temperatures should also be avoided to prevent the induction of cherry crinkle disease. High summer temperatures (above 35°C) also are associated with fruit doubling the following season. Bing is extremely susceptible to fruit doubling, whereas Jubilee, Rainier, Brooks, Tulare, and King produce fewer doubles under similar conditions. The ideal soil is a well-drained, deep gravelly or sandy loam. Cherries tend to be sensitive to excess amounts of boron, chloride, sodium, and total salts in the irrigation water. Under current planting designs, sweet cherries require less pruning than other stone fruits due to the persistent nature of sweet cherry fruiting spurs. Annual moderate pruning is used to renew 10–30% of the fruit-bearing wood. Fruit destined for high-quality fresh markets should be grown on trees that are pruned more heavily to control excessive fruit, thereby producing larger, sweeter fruit.

Harvest and Handling

Processing cherries can be harvested mechanically, whereas those for fresh market are harvested by hand. Mechanical harvesting is accomplished with the use of tree shakers. Fruit for fresh market are picked with the pedicel intact to minimize the potential for fungal decay. Cherries are picked as ripe as possible because sugar does not increase after harvest. After harvest, cherries are held at low storage temperatures, as close to 0°C as possible, to delay softening of the fruit and the progress of decay. High relative humidity is an additional storage requirement to prevent or minimize moisture loss. Approximately 70–90% of the sweet cherries are sold for fresh market, and the remainder are used for processing. They are shipped in international trade, mainly from Chile to the United States, and from the United States to Japan. Those used for processing are either for brining for maraschino cherries or for canning or freezing.

Selected References

Cherry Marketing Institute. 1992. Cherries and cherry processing. Pages 369-374 in: Encyclopedia of Food Science and Technology, vol. 1. Y. H. Hui, ed. John Wiley & Sons, New York.

Gardner, V. R. 1946. The Cherry and Its Culture. Orange Judd Publishing Co., New York. 146 pp.

Micke, W. C., and Mitchell, F. G. 1972. Handling sweet cherries for the fresh market. Univ. Calif. Agric. Exp. Stn. Ext. Serv. Circ. 560. 18 pp.

Norton, R. A., Hansen, C. J., O'Reilly, H. J., and Hart, W. H. 1963. Rootstocks for sweet cherries in California. Calif. Agric. Exp. Stn. Bull. 159. 5 pp.

Westwood, M. N. 1978. Temperate Zone Pomology. W. H. Freeman & Co., San Francisco. 428 pp.

(Prepared by P. R. Rener and S. M. Southwick)

Sour Cherry

Origin and Production

Sour cherry (*Prunus cerasus* L.), also called tart cherry, is believed to be an interspecific hybrid between sweet cherry (*P. avium* L.) and ground cherry (*P. fruticosa* Pall.). Sour and ground cherry are tetraploids with a haploid number of 16 chromosomes, while sweet cherry is diploid. According to Vavilov, sour cherry arose in the Near East, which includes portions of Asia Minor, Iran, Iraq, and Syria, but Hedrick believed the center of origin also included Switzerland and the Adriatic Sea. Ground cherry exists in great diversity in Russia, especially in the Volga region, the Ural Mountains to the 60th parallel, and southwestern Siberia. Sweet cherry grows wild throughout the temperate regions of mainland Europe and southern Russia, whereas the habitat of sour cherry in eastern Europe and Asia Minor overlaps with sweet cherry in the south and ground cherry in the north. Hybridization between sweet and ground cherry may have occurred many times, and the three species continue to hybridize in regions where they coexist. Tetraploid hybrids between sour cherry and sweet cherry are commonly called Duke cherries. Other "sour" cherry cultivars are natural hybrids between sour cherry, ground cherry, Duke cherries, and sour × ground cherry hybrids crossed again with sour cherry. As a result of this interspecific hybridization, sour cherry germ plasm is extremely diverse in characteristics inherited from the two progenitor species.

The leading countries in production are Russia (33.6%), Turkey (15.5%), Germany (12.4%), Serbia (11.2%), the United States (9.9%), Poland (6.6%), Hungary (5.3%), other Western (2.7%) and Eastern (2.2%) European countries, and Canada (0.6%). In the United States, most sour cherries are produced in Michigan, New York, Utah, and Wisconsin, with limited production in Pennsylvania, Oregon, and Washington.

Botany and Biology

Sour cherry produces vegetative buds, which develop into new shoots, and flower buds, which usually have two to five flowers within each bud. A vegetative bud develops either into a vigorous shoot or a spur, which is a lateral or short side shoot that grows less than 2.5 cm per season. Spurs typically have a terminal leaf bud and a cluster of lateral blossom buds. Vigorous shoots normally develop only leaf buds at the terminal end of the shoot and blossom buds on the basal end. These flower buds (called lateral flower buds as opposed to the spur flower buds) will fruit the following year on this one-year-old wood, but the site is unproductive in later years.

The commonly grown cultivar Montmorency cherry typically produces 70% of its fruit from flower buds on one-year-old wood and 30% from spur flower buds, but percentages depend on tree vigor. Healthy trees have a large proportion of moderately vigorous shoots that form leaf buds, which will develop into fruiting spurs. Although most cultivars, such as Montmorency, fruit on one-year-old wood and spurs, certain cultivars fruit exclusively on one or the other. For example, the Hungarian cultivar Erdi Nagygyumolcsu produces all fruit on spurs, whereas the Serbian cultivar Sumadinka produces all fruit on one-year-old wood.

Tree height ranges among cultivars from 10 m (similar to sweet cherry) to dwarfs of 1 m, similar to ground cherry. Sour cherry cultivars may be self-compatible, partially self-compatible, or self-incompatible. Self-incompatible trees must be planted with a pollinator variety. Self-incompatibility in sour cherry resembles gametophytic self-incompatibility, with complete inhibition of self-pollen tube growth in the style. The inheritance of self-incompatibility in sour cherry is not known.

Horticulture

Cultivars

Sour cherry production in the United States is a monoculture of the 400-year-old cultivar Montmorency. In Eastern Europe, the center of diversity for sour cherry, the numerous cultivars grown span a ripening period of 40 days. In most cases, the parents of the Eastern European varieties are local selections propagated for centuries by peasants.

Sour cherry cultivars sometimes are classified into two subgroupings based on flesh and juice color: morello (dark-fleshed with red juice) and amarelle (light-fleshed with colorless juice). However, sour cherries exhibit a range of juice and flesh color from clear through shades of pink, scarlet, red, and purple. The major cultivar in northern Europe is the dark-fleshed Schattenmorelle.

Rootstocks

Most sour cherry cultivars are budded onto open-pollinated seedlings of *P. mahaleb* or *P. avium.* Size-controlling rootstocks are not needed for sour cherry, because small-statured sour cherry cultivars exist, and sour cherry can be mechanically harvested with trunk shakers. In Romania, sour cherry selections that are incompatible with *P. mahaleb* and *P. avium* are grown on sour cherry seedlings. Several breeding programs have introduced clonal sour cherry rootstocks, but none have become commercially accepted.

Growing Systems, Harvest, and Processing

In the United States, cultivar Montmorency is typically planted at a spacing of 5 × 6 m, and fruit is harvested with trunk shakers. In many countries, however, much of the crop is hand-harvested, and cultivars of low vigor are grown. In Romania, low-vigor cultivars are spaced at 2.5 × 4 m and trained to a palmette system with three wires. Oblacinska, the major cultivar in Serbia, is grown on its own roots and planted at a spacing of about 3 × 5 m.

Most Montmorency production in the United States is pitted mechanically for use in pies. In Europe, cultivars with different fruit characteristics are grown for different products. For example, early-maturing cultivars with large fruit size are sold on the fresh market, whereas cultivars with dark (red-purple) flesh are used for juice. These diverse fruit characteristics offer a wealth of variation for utilization by the plant breeder and ultimately for benefit to the consumer.

Selected References

Apostol, J., and Iezzoni, A. 1992. Sour cherry breeding and production in Hungary. Fruit Var. J. 46(1):11-15.

Hedrick, U. P. 1915. The Cherries of New York. J. B. Lyon Company, Albany, NY. 371 pp.

Vavilov, N. I. 1951. The Origin, Variation, Immunity and Breeding of Cultivated Plants. Ronald Press, New York. 364 pp.

(Prepared by A. F. Iezzoni)

Part I. Infectious Diseases

Diseases Caused by Fungi

Fruit Diseases

Brown Rot

Brown rot, a major disease of all commercially grown *Prunus* species in most regions of the world, is caused by three closely related fungi—*Monilinia fructicola* (G. Wint.) Honey, *M. laxa* (Aderhold & Ruhland) Honey, and *M. fructigena* Honey in Whetzel. The distributions of the pathogens overlap to some extent, but *M. fructigena* was eliminated from North America, and *M. fructicola* has not been detected in Europe. *M. fructigena* occurs on pome and stone fruits in Europe, but crop losses are less there than those caused by *M. laxa*, which is found in all major fruit production countries. In Europe, South Africa, Chile, and Iraq, *M. laxa* is the most common brown rot pathogen. *M. laxa* causes serious blossom blight and fruit decay problems on apricot, prune, and sometimes sweet cherry, nectarine, and peach. *M. laxa* is more widespread in California than it is in midwestern and northeastern states. It is not found in the southeastern United States. *M. fructicola* was identified first in the eastern United States and has caused serious crop losses in this country and also in Canada, Australia, New Zealand, Brazil, and Japan. In the eastern United States, *M. fructicola* has the potential to cause severe crop losses year after year, and in California losses may be severe when weather favors infection of green fruit or when unexpected rains occur during harvesttime.

Financial losses caused by brown rot include direct crop loss from blossom and twig blight and fruit rot as well as the expense of disease control. In California in the early 1950s, complete destruction of ripe peach fruit followed a single rain, but losses of such magnitude have been rare in recent times. Improved management and more effective fungicides are responsible for the reduced damage.

Symptoms

Typical disease symptoms induced by *Monilinia* spp. are blossom and twig blight (Plate 1) and fruit rot (Plate 2).

Blossom and Twig Blight—In spring, the primary features of brown rot caused by *M. fructicola* and *M. laxa* are blossom blight, twig blight, and green fruit rot. Blossom blight on peach is not common in the eastern United States, except in the Southeast, and is caused only by *M. fructicola*. In the western United States, blossom blight on peach is caused by both pathogens. On peach, the first symptom of blossom blight is necrosis of the anthers, which proceeds to the floral tube, ovary, and peduncle. Infections may extend into and then girdle the twig. As infected flowers wilt and turn brown, they often become affixed to the twig in a gummy mass. In wet weather, infected flowers become covered with grayish to tan sporodochia.

On apricot, plum, and prune, all blossom parts are susceptible. Apricot is most susceptible to blossom blight, followed in order by prune, sweet cherry, peach, sour cherry, and plum. Symptoms appear first on sepals, then on petals, anthers, and stigmata as the blossoms open. Blight symptoms are expressed when the fungus girdles the peduncle. By this time, sporulation of *Monilinia* is usually evident on infected flower parts but not on flowers blighted from girdling of twigs. These other blighted blossoms may become infected with *Cladosporium*, *Botrytis*, *Alternaria*, or other fungi.

On sour cherry in the Great Lakes region, *M. laxa* or *M. fructicola* can cause blossom and spur blight. Conidia form on dead spurs, twigs, and cankers. Elsewhere in the Midwest and eastern United States, *M. fructicola* predominates. Blossom blight on sweet cherry also is caused by both *M. fructicola* and *M. laxa*. Blossom parts attacked are stigma, anthers, and petals, and the fungus moves into the floral tube and pedicel and occasionally into the spur. Profuse gumming usually occurs on twigs at the margin between diseased and healthy tissue on apricot, plum, and prune, and sometimes on cherry, nectarine, and peach.

Reasons for differences in pathogen specificity between *M. laxa* and *M. fructicola* on hosts in various localities are not well understood but could be related to temperature and humidity. Since the introduction of the benzimidazole and dicarboximide fungicides, occurrence of populations of these two species could be related to their sensitivity to fungicides used.

Cankers—Frequently, following *M. laxa* colonization of the blossom, the fungus enters the twig or branch, where it causes an elliptical-fusoid canker with massive gum formation at the advancing margin. At times, the cankered area girdles the twig and blights the portion distal to the cankered area. Leaves on such shoots turn tan to brown and remain attached instead of abscising. Usually cankers are restricted to twigs and do not extend into the previous year's wood. Blighted blossoms or twigs are almost invariably evident in the center of the branch cankers. Cankers do not continue to enlarge from one season to the next. If the branch is not girdled, surrounding healthy tissue will produce callus. Sporulation may continue on large cankers for four years or more. Cankers and killed shoots may be colonized by other more aggressive canker-causing fungi such as *Leucostoma* species. In California, large cankers on twigs or branches from blossom infections caused by *M. fructicola* are not common in apricot, prune, and peach orchards even with severe blossom and twig blight, whereas in the eastern United States, cankers induced by *M. fructicola* are common in some peach orchards.

Leaf Infection—The occasional infection of young leaves

by *M. laxa* can cause blighting of terminal shoots. Infections on apricot leaves induce small, red lesions, which are similar to but much smaller than those caused by the shot-hole pathogen, *Wilsonomyces carpophilus*.

Fruit Infection—Brown rot on ripening or mature fruit typically develops as a rapidly spreading, firm, brown decay. Under optimum conditions, decay of ripe peaches infected by *M. fructicola* may be visible within 48 hr of infection, but in drier locations, 72 hr may be required. In some areas, notably Australia and California, infection of fruitlets during or shortly after flowering may result in quiescent infections on green fruit that become active before or after harvest. Quiescent infections have not been reported in the eastern United States except on plum, perhaps because blossom infection is less common there than in some other locations. Quiescent infections may remain invisible or appear as small, circular, necrotic lesions (Plate 3) that enlarge rapidly as fruit ripens. On immature sweet cherry fruit, symptoms of quiescent infections range from superficial brown to red halos, 0.5–2 mm in diameter, to necrotic pitted areas up to 7 mm in diameter. As stone fruit mature, decay spreads from these initial lesions to involve large areas of the fruit and may spread by contact to adjacent fruit.

Immature fruit that wither on the tree after failure of pollination or following injury may become infected and covered with conidia. Likewise, green fruit dropped on the orchard floor during thinning may be infected. Otherwise, immature fruit are seldom infected unless they are in contact with diseased tissue or injured by frost or insects. Immature, mummified fruit on which *Monilinia* species sporulate profusely are seldom observed in California orchards, but in Brazil and the eastern United States and Canada they are more common.

Usually, both immature and mature fruit with brown rot tend to remain attached to the tree as toxins from the decaying tissue cause shoot dieback and prevent formation of an abscission layer between fruit and peduncle. With both *M. laxa* and *M. fructicola*, the sporulating area, which is mouse gray to tan, will have concentric zones of sporulation produced in response

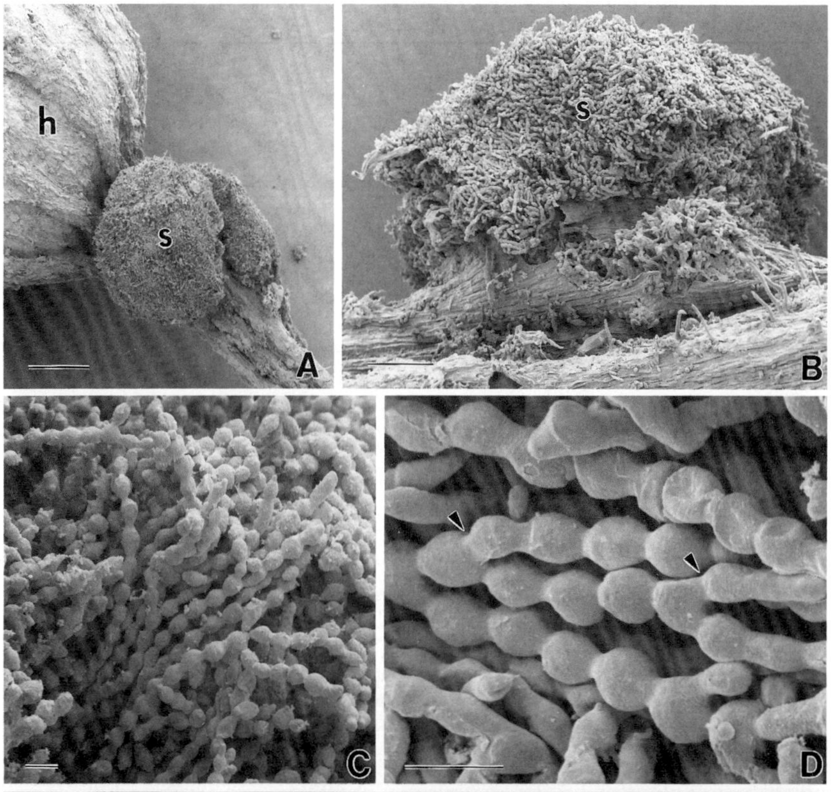

Fig. 1. Scanning electron micrographs of *M. laxa*. **A–E**, Specimens hydrated, fixed in glutaraldehyde, and dehydrated in alcohol specimen preparation: **A**, Sporodochia (s) on peduncle of blighted blossom below hypanthium (h); bar = 400 μm. **B**, Lateral view of "cushion-shaped" sporodochium (s); bar = 100 μm. **C**, Masses of conidial chains in a sporodochium; bar = 10 μm. **D**, Branching points of conidial chains (arrowheads); bar = 10 μm. **E**, Hydrated spore chain; bar = 100 μm. **F**, Nonhydrated and freeze-dried specimen preparation; in situ conidial chains showing partial collapse of conidia with ellipsoidal conidial initials forming in acropetal succession; bar = 10 μm. (Courtesy J. E. Adaskaveg and M. Dunlap)

to diurnal cycles. With *M. fructigena,* sporodochia are larger, scattered, and yellow-orange. Often, decaying fruit in cold storage will appear black with little or no sporulation. The black areas (pseudosclerotia) consist of both fungal and host tissues.

Causal Organisms

In 1928, Honey erected the genus *Monilinia* for the monilioid species previously placed in the genus *Sclerotinia.* The generic name for the conidial stage is *Monilia* (Fig. 1). Conidia are produced in sporodochia in branched, monilioid chains without disjunctors. Species identification by conidial morphology is somewhat difficult, because the hyaline, lemon-shaped conidia are similar: 8–23 × 7–16 µm for the summer (winter 5–19 × 4–12 µm) spores of *M. laxa,* 8–28 × 6–19 µm for those of *M. fructicola,* and 12–34 × 9–15 µm for *M. fructigena.* The spore pustule (sporodochium) appears gray at first but later becomes light tan. Microconidia (2.5–3 µm in diameter), which are borne on bottle-shaped, often asymmetric phialides, are present on decaying fruit infected with *M. fructicola* and in old cultures of *M. fructicola* on synthetic media. These microconidia seldom germinate, and their role as infective propagules has not been established. Some believe that microconidia may function as spermatia to produce apothecia.

Apothecia of *M. fructicola* (Plate 4) can develop from fruit mummies partially buried in soil or debris on the orchard floor at the same time the host blossoms. Apothecia of *M. fructicola* are light brown, about 5–20 mm in diameter at maturity, cupulate, and stipitate. Asci are 102–215 × 3–13 µm and ascospores 6–15 × 4–8 µm. Apothecia of *M. fructigena* and *M. laxa* are rare in the field and have not been produced in culture.

Species cannot be identified solely by symptoms and signs on infected tissues. Growth characteristics on 2% potato-dextrose agar under diurnal light conditions are useful for distinguishing *M. laxa* from *M. fructicola* (Plate 5). Whereas *M. fructicola* grows rapidly, with concentric rings of many spores, *M. laxa* grows more slowly, and the colonies are lobed with few spores. Cultures of both fungi grown under continuous light sporulate profusely, especially on tomato juice or oatmeal agar and on drained canned peach, apricot, or pear fruit pieces. Another method for differentiation is to inoculate mature green pear fruit. *M. fructicola* sporulates abundantly, whereas *M. laxa* sporulation is sparse. A third useful characteristic is that interaction (barrage) zones have not been observed between colonies of *M. laxa* isolates, but they occur when isolates of *M. laxa* and *M. fructicola* are paired.

Additional taxonomic techniques for species differentiation include electrophoresis, isoelectric focusing, and serology.

Disease Cycle and Epidemiology

Monilinia species overwinter on treeborne mummies, peduncles, blighted blossoms and twigs, and cankers. Sporodochia (Fig. 1A and B) develop on blighted blossoms and twigs under cool, wet conditions during the winter and early spring. At times, sporodochia may be found on prunings on the orchard floor, where infected tissues are exposed to moisture for a longer period. Conidia are generally formed on fruit mummies and blighted twigs and blossoms at temperatures of 5°C or higher. Conidia are disseminated by wind and rain and germinate rapidly under favorable conditions. The incidence of blossom blight caused by *M. fructicola* is related to temperature and duration of wetness; as little as 3–5 hr of wetting at 20°C may lead to significant infection. With 24 hr of wetting, severe blossom infection may result regardless of temperature. Optimum temperature for blossom infection of peach and cherry is 25°C. In general, within the range of 5–30°C, temperatures above or below the optimum delay infection but do not prevent it. For *M. laxa,* infection and development of blossom and twig blight occur during periods of wetness at temperatures above 13°C (optimum 24°C). Inoculum concentration also interacts with temperature and wetness duration to influence the incubation period, disease incidence, and severity. Conidia produced on mummies or blighted blossoms remain viable throughout the hot, dry summer months in California.

Apothecia of *M. fructicola* form on mummies on the orchard floor (Plate 4), but are not commonly produced in most commercial orchards. Buried or partially buried mummies exposed to continuous moisture are most likely to produce apothecia. Also, mummies in weeds under trees or in stands of wild plums, which are common near peach orchards in the southeastern United States, produce apothecia more frequently than mummies under exposed, dry conditions. Apothecia have never been observed on mummies hanging on the tree. In years when *M. fructicola* apothecia are common, subsequent severe blossom blight has been noted in peach and apricot orchards. Nevertheless, conidia produced on mummies and cankers are the more common inoculum for blossom infection. Although *M. fructicola* seldom infects flowers to the extent that the crop is reduced, conidia produced on blighted blossoms and shoots provide inoculum for fruit infection, and invasion of shoots provides for survival of the pathogen in twigs from one season to the next. Sporodochia of *M. fructicola* on infected blossoms and twigs may produce viable conidia throughout the remainder of the current growing season, and these conidia may infect injured green fruit; nonabscised, aborted fruit; green fruit thinned after the pit hardening stage and dropped under the trees; or ripening fruit as they mature. Susceptibility of stone fruit to infection increases as fruit color begins to develop. Fruit infection occurs by direct penetration through the cuticle, stomata, and base of trichomes as well as through suture cracks or injuries.

Insects (nitidulid beetles, vinegar flies, and honeybees) can be important vectors of *M. fructicola* during fruit ripening, carrying conidia to injuries produced by oriental fruit moths, peach twig borers, Mediterranean fruit flies, plum weevils, or even cucumber beetles, bees, or ants. Wounded fruit are infected more readily than nonwounded fruit. At harvest, apparently healthy fruit usually are contaminated with spores and/or quiescent infections (Plate 3), and decay may occur during storage and marketing.

Control

Protective fungicidal treatments provide the best control for both blossom blight and fruit rot. Without postharvest treatment, fruit usually cannot be stored safely at room temperature. The proper use of fungicides with some systemic activity protects flowers and fruit, reduces the amount of sporulation formed on the infected tissue, and reduces sources of overwintering inoculum.

Fungicides available for brown rot control in the United States are the dicarboximides (iprodione and vinclozolin), benzimidazoles (benomyl and thiophanate methyl), triforine, chlorothalonil, myclobutanil, fenbuconazole, and propiconazole. Other fungicides used are copper compounds, sulfur, and captan. The selection of a fungicide or mixture of compounds often is influenced by the need to control other diseases that may occur more or less simultaneously with brown rot, such as scab, powdery mildew, rust, russet scab, or gray mold. Timing of application is critical for blossom blight control; fungicides must be applied when susceptible flower parts are exposed and before or soon after the occurrence of periods of wetness and temperatures that are conducive to infection. Fungicides need not be applied to immature fruit unless moisture conditions are especially favorable (eastern United States) for infection, or injury caused by insects or cold has increased the likelihood of decay. Thus, insect control during this period may be an important consideration.

Wetness of fruit induced by dew or sprinkler irrigation can trigger infection by the brown rot organisms. To reduce the number of fungicide applications, a forecasting method that considers such parameters as wetness period, temperature, amount of available *M. fructicola* inoculum, and fruit maturity is being developed for arid regions where summer rains are rare. A single rain shower during fruit maturation can cause devastating losses in California, whereas losses tend to be less severe in regions (eastern United States, Brazil, Japan) that normally have multiple rain showers during this period but where multiple fungicide applications usually are made.

Fungicide-resistant *M. fructicola* and *M. laxa* became evident following multiple applications of benomyl over a five-year period. Resistance has not been commonly experienced with other fungicides, including the more recently introduced dicarboximides and demethylation inhibitors (e.g., iprodione, vinclozolin, and triforine) in the United States. Yet, in New Zealand, resistance to dicarboximides has been shown, but crop losses have not been experienced. Benomyl-resistant strains of *M. fructicola* are stable and competitive with benomyl-sensitive strains. To reduce or prevent selection of dicarboximide-resistant strains, the recommendation is to reduce the number of applications.

Cultural practices such as removal of mummies and pruning infected twigs reduce the inoculum level, but these procedures alone are not sufficient to control the disease. Attempts to reduce apothecial development by orchard cultivation are not effective. Practices that reduce stress through good nutrition or proper soil moisture can increase tolerance to blossom infection, but the benefit is marginal.

Control of insects that serve as vectors and/or provide wounds for infection is essential for effective brown rot control. Fungicides do not control infection if applied following inoculation of a mechanical injury. In California, control of the oriental fruit moth and peach twig borer reduces feeding sites for spore-contaminated nitidulid beetles. In Brazil, where the Mexican or Mediterranean fruit fly is a problem, effective brown rot control is difficult to obtain. In the eastern United States, control of the plum curculio is essential for preventing brown rot.

Nonfungicidal treatments under experimentation include materials that induce fruit resistance, such as calcium treatments or surface coatings. The Bolinha peach cultivar in Brazil has a resistance mechanism in the epidermal layer that results in a longer incubation period for infection than that for the more susceptible cultivars grown in the United States.

To reduce postharvest decay, rapid cooling (hydrocooling or cold storage) of the fruit delays development of brown rot. Artificial ripening at high temperature (35°C, which is too warm for the growth of *M. fructicola*) is used in Australia. Effective control of *Monilinia* and *Rhizopus* postharvest decay pathogens has been obtained with a spray mixture of iprodione with certain wax-oil emulsion combinations.

Selected References

Adaskaveg, J. E., Ogawa, J. M., and Feliciano, A. J. 1992. Comparison of calcium-based and film-forming materials for control of brown rot of peach caused by *Monilinia fructicola*. (Abstr.) Phytopathology 82:1158.

Biggs, A. R., and Northover, J. 1988. Early and late-season susceptibility of peach fruits to *Monilinia fructicola*. Plant Dis. 72:1070-1074.

Biggs, A. R., and Northover, J. 1988. Influence of temperature and wetness duration on infection of peach and sweet cherry fruits by *Monilinia fructicola*. Phytopathology 78:1352-1356.

Byrde, R. J. W., and Willetts, H. J. 1977. The Brown Rot Fungi of Fruit: Their Biology and Control. Pergamon Press, New York. 171 pp.

Ogawa, J. M., and English, H. 1991. Diseases of Temperate Zone Tree Fruit and Nut Crops. Publ. 3345. University of California, Division of Agriculture and Natural Resources, Oakland. 461 pp.

Weaver, L. O. 1950. Effect of temperature and relative humidity on occurrence of blossom blight of stone fruits. Phytopathology 40:1136-1153.

Wilcox, W. F. 1989. Influence of environment and inoculum density on the incidence of brown rot blossom blight of sour cherry. Phytopathology 79:530-534.

(Prepared by J. M. Ogawa, E. I. Zehr, and A. R. Biggs)

Shot Hole

Shot-hole disease, caused by *Wilsonomyces carpophilus* (Lév.) Adaskaveg, Ogawa, and Butler, can be severe on apricot, peach (Plate 6), and nectarine but is of no consequence on plum or prune. It sometimes causes damage to sweet cherry in the Pacific Northwest but is rare on this host in California. The geographic range of this disease includes Australia, Africa, Asia, New Zealand, Europe, and North and South America. Severe outbreaks of shot hole have been experienced following years of economic depression, when dormant sprays were not applied and the winter phase of the disease was allowed to progress.

Symptoms

Symptoms on peach, nectarine, and apricot are similar. During wet winter months, the fungus infects and kills dormant buds. Dead buds sometimes are covered with a gummy exudate that gives them a varnished appearance (Plate 7). Stem lesions, mostly 3–10 mm in diameter with sporodochia in the center, are common on peach and nectarine but not on apricot. Leaf and fruit lesions begin as small purplish areas that expand to brownish spots 3–10 mm in diameter. On fruit, lesions develop mostly on the upper side and eventually become corky and rough. Leaf lesions abscise in a warm, dry environment; thus the name "shot hole." Such early dehiscence largely eliminates the formation of sporodochia on the lesions. Defoliation after infection is not common on peach and nectarine but is of some importance. Blossom infection has not been observed on stone fruit, but it does occur on almond.

Causal Organism

The genus *Wilsonomyces,* erected for the shot-hole fungus, is characterized by sporodochial conidiomata, sympodially developing conidiogenous cells, and smooth-walled phragmospores (multicelled) that separate from the conidiogenous cell by circumscissile splitting of the wall of the cell below the conidium (rhexolysis). Previous generic names for this fungus include *Clasterosporium*, *Coryneum*, and *Stigmina.* The characteristic conidia are solitary, unbranched, thick-walled, and ellipsoidal or fusiform, with three to five transverse (occasionally oblique), double-walled septa (Plate 8). Dimensions are in the range of 20–90 × 7–16 μm (3.5 μm at the base). Conidia are subhyaline to golden brown but appear olive to black in mass. Conidia are mass-produced in sporodochia that are usually punctiform, tan to olivaceous brown and then black, and formed on twigs, leaves, and fruit (infrequently) of *Prunus* species.

Isolation of the pathogen is more easily accomplished from leaf than from fruit lesions. For identification, sporulating lesions are crushed in water, and the conidia are observed under the microscope. To establish pure cultures, a single conidium can be placed on potato-dextrose agar. Cultures will sporulate in light within a week at 20°C.

Disease Cycle and Epidemiology

Disease development on peach, nectarine, and apricot is similar. During wet winter months, the fungus sporulates in infected buds and on twig lesions of the previous season. Conidia are not easily detached from the conidiophores by moving air but are readily removed by water. Dispersed conidia can remain viable for several months on the twigs. On potato-dextrose agar at room temperature, germ tubes can emerge from conidia within 1 hr. After wetting, only one or two cells of a conidium may produce germ tubes, but other cells may be activated upon rewetting. Twig infection requires at least 24 hr of continuous wetness and can occur at winter temperatures where most stone fruit are grown. Direct penetration follows formation of an appressorium covered with a gelatinous matrix. The incubation period is five to 14 days, depending on temperature and the type of tissue infected. Mycelial growth in vitro occurs at 4–30°C, with an optimum temperature range of 15–20°C. Spores can germinate at temperatures as low as 1°C.

Control

Control of shot-hole disease on peach and nectarine requires protection of dormant buds, shoots, and fruit, whereas on apricot, fruit and leaves must be protected as well as dormant buds. A single dormant spray of Bordeaux mixture or a fixed copper formulation applied before autumn rains provides protection throughout the winter. To control fruit and leaf infection (primarily on apricot), a spray application immediately after fruit set is common. Because copper compounds can be injurious to young leaves, organic fungicides such as captan, ziram, iprodione, or chlorothalonil are used for this purpose.

Ground sprayer or aircraft applications can be used during the dormant stage, but ground sprayers are preferred after leaves emerge. No commercially acceptable cultivars of apricot, peach, or nectarine with resistance are available. Sprinkler irrigation that wets leaves and fruit can increase disease incidence.

Once the fungus is well established in perennial infections within a tree, effective disease control with protective fungicides is difficult. Pruning diseased wood is the only available practice to eradicate or reduce the amount of fungal inoculum, but this is often impractical.

Selected References

Adaskaveg, J. E., Ogawa, J. M., and Butler, E. E. 1990. Morphology and ontogeny of conidia in *Wilsonomyces carpophilus*, gen. nov. and comb. nov., causal pathogen of shothole disease of *Prunus* species. Mycotaxon 37:275-290.

Ogawa, J. M., and English, H. 1991. Diseases of Temperate Zone Tree Fruit and Nut Crops. Publ. 3345. University of California, Division of Agriculture and Natural Resources, Oakland. 461 pp.

Wilson, E. E. 1937. The shot-hole disease of stone-fruit trees. Calif. Univ. Agric. Exp. Stn. Bull. 608:3-40.

(Prepared by J. M. Ogawa and H. English)

Scab

Scab is an important disease of peaches and nectarines in the warm, humid peach-producing areas of the world. The disease also occurs rarely on plums and apricots. A related fungus causes a scab disease on cherries, but this disease is not economically important. In one report, severe scab was found on sour cherry but only light infection on sweet cherry.

Symptoms

Peaches and Nectarines—Twigs, leaves, and fruit are infected, but the symptoms are most noticeable on fruit. Symptoms appear when the fruit are about half grown. Small (less than 0.5 mm), greenish to olivaceous, circular spots appear on exposed fruit surfaces, most frequently near the stem end. These enlarge to 2–3 mm, become olivaceous to black, and sometimes have a green to yellow halo on the blush surface (Plate 9). Lesions may be larger on nectarines. When numerous lesions coalesce, cracking occurs as the fruit enlarges. Cracked fruit are more susceptible to brown rot. Lesions per fruit can be very numerous.

Twig infection occurs on tender, green twigs of new growth, and some lesions appear simultaneously with fruit lesions. Initially, twig infections are barely visible. They are slightly raised, circular to oval and the same color as surrounding normal tissue. With time, they become brown with a slightly raised border (Plate 10). By the end of the growing season, the border becomes purple to dark brown, and the lesion measures 3–5 × 5–8 mm and is slightly elongated along the stem. Although some sporulation may occur on developing twig lesions, heavy sporulation usually occurs on olivaceous tufts of conidiophores during the next season. Many twig lesions will not be apparent during the season when infection occurs but will be visible the next spring. The fungus rarely survives the second winter on twigs. The lesions usually do not cause discernible damage to the twigs, except in nursery stock where dieback may occur if the disease is not controlled.

Young or mature leaves are infected on the lower surface (Plate 11). Initially, lesions are imperfectly defined, angular to circular, and of a color similar to that of healthy tissue. Lesions turn olive green when the fungus begins to sporulate. Fully developed lesions are usually 1–2 mm in diameter but may be larger. In severe disease situations, lesions may coalesce, causing chlorotic areas on the leaf that become necrotic. Natural leaf infection seldom causes defoliation sufficient to reduce vigor of the trees.

Sour Cherries—The first evidence of fruit infection is small, pink splotches. Lesions are flat, bright red around the periphery, and dull red to brown in the center; they are not well demarcated from surrounding tissue, and enlarge slowly.

Causal Organism

Cladosporium carpophilum Thuem. (teleomorph, *Venturia carpophila* E.E. Fisher) forms stromatic masses in culture and on the host. Young hyphae are hyaline, but older hyphae develop thickened, exposed walls and become olivaceous. Conidiophores are short and erect, are rarely branched, have one or more septa, and are enlarged at the base. When developing from overwintering stromatic masses on twigs, they occur in tufts. Conidia are formed at the apex of the conidiophores and may be borne singly or in chains. Conidia are one- or two-celled. If two-celled, they measure 4.2–5.2 × 14.9–17.2 µm and are constricted at the septum. Chlamydospores are formed in culture and on twigs in the field.

The fungus grows well on most media. The optimum temperature for growth in culture is 20–25°C; growth is slow at 10 and 32°C, and no growth occurs at 35°C.

Venturia cerasi Aderhold is the causal organism of cherry scab. The fungus is widespread in Europe, Brazil, Canada, Iran, and New Zealand, but the last reported outbreak of cherry scab occurred in East Malling, England, in 1943. Although little research has been done on this organism, the life cycle appears to be the same as or similar to that of *C. carpophilum*.

Disease Cycle and Epidemiology

C. carpophilum overwinters as mycelia in lesions on twigs and as chlamydospores on the bark surface. Conidia are produced first about two weeks before calyx split and increase in numbers until three to four weeks after calyx split. Viability of conidia decreases rapidly with age, but fresh conidia are produced rapidly when weather favors infection. Sporulation on twigs occurs at 70–100% relative humidity (RH) and increases with humidity. Sporulation may occur in as little as 3 hr at 100% RH.

Although conidia germinate at 15–30°C, the optimum temperature for germination is 25–30°C. Conidial germination occurs at 94–100% RH and is optimized at 98–100% RH.

Liberation of conidia from lesions on infected twigs decreases as RH decreases from 100 to 40%. Further decreases in RH with exposure to infrared radiation result in increased spore discharge.

Fruit infection seldom occurs in nature until 30 days after petal fall, even though fruit are susceptible much earlier. Before this time, young peach fruit have abundant hairs that discourage fungal penetration and are positioned such that conidia do not wash over them from subtended twigs. Nectarines, lacking pubescence, presumably are susceptible to invasion earlier, but experimental evidence is lacking. Conidia in the air are in greatest numbers from two to six weeks after calyx split, which is considered to be the period of greatest danger of fruit infection. Twig infection can occur whenever new growth is present, weather conditions are favorable, and spores are being produced. Because little shoot growth is present until after calyx split, twig infection probably occurs later than fruit infection.

The fungus develops in close association with the host epidermal cells. On fruit, penetration occurs directly at the base of hairs on peach. Cork layers form under the infected area and compartmentalize the fungus after the superficial host cells die. In early-maturing cultivars, the fruit may mature and be harvested before cork formation occurs. Similar development occurs on twigs. Early-season twig infections, which sporulate during summer, are walled off by a periderm. During winter, the fungus grows beyond the periderm to form a secondary lesion, which appears as an elevated ring of dark red tissue surrounding the primary lesion. Late infections form only primary lesions. Abundant sporulation occurs on both types of lesions in spring. As the bark forms and roughens during the second year of growth, the cuticle sloughs, and the lesions lose their identity.

The period of latent infection on fruit is 42–77 days. On leaves, it varies from 25 to 45 days. On twigs, symptoms may appear after 25 days, but many lesions do not become visible until the next spring.

There have been no reports of the role of the teleomorph stage.

Control

Scab on peaches, nectarines, and apricots is controlled entirely by fungicidal sprays. Sanitation is impractical because numerous overwintering lesions persist on fruiting wood.

Pruning trees in a manner that allows good penetration of sunlight and unimpeded air movement can help in scab control by facilitating rapid drying and good spray penetration.

Usually, fungicides should be applied at calyx split and every two weeks thereafter for a total of four or five sprays. More frequent applications may be needed in humid climates and fewer applications in cooler areas. Wettable sulfur, captan, benzimidazoles, and chlorothalonil have given excellent control of scab, but resistance to benzimidazoles prevails in some areas.

Selected References

Bensaude, M., and Keitt, G. W. 1928. Comparative studies of certain Cladosporium diseases of stone fruits. Phytopathology 18:313-329.

Gottwald, T. R. 1983. Factors affecting spore liberation by *Cladosporium carpophilum*. Phytopathology 73:1500-1505.

Keitt, G. W. 1917. Peach scab and its control. U.S. Dep. Agric. Bull. 395. 65 pp.

Koch, L. W. 1934. Studies on the overwintering of certain fungi parasitic and saprophytic on fruit trees. Can. J. Res. 11:190-206.

Lawrence, E. G., and Zehr, E. I. 1982. Environmental effects on the development and dissemination of *Cladosporium carpophilum* on peach. Phytopathology 72:773-776.

Moore, M. H. 1944. Cherry scab (*Fusicladium cerasi* (Rabenh.) Sacc.) in Kent in 1943. Pages 54-56 in: Rep. East Malling Res. Stn.

(Prepared by F. F. Hendrix, Jr.)

Powdery Mildew

Apricot, cherry, nectarine, peach, and plum are susceptible to powdery mildew. In orchards, the largest economic losses usually result from fruit infections, but foliar mildew is most damaging in nursery plantings. The disease on the various hosts occurs over a wide geographic area and is particularly troublesome in the semiarid areas of California, the Pacific Northwest, and Eastern Europe.

Crop losses resulting from fruit infections may reach 100% on cherries to 50% on Japanese plums, apricots, nectarines, and peaches, but none on French prunes. Injurious effects on tree growth and crop production from foliage, stem, and bud infections are unknown, but effects on nursery trees may be significant. Protective sprays on cherries, plums, and peaches are costly, because multiple applications usually are necessary. On tart cherries, pedicel infections may become so severe that mechanical harvesting is sometimes impeded.

Symptoms and Signs

With the exception of cherry and plum, young leaves of stone fruit are most susceptible to *Sphaerotheca pannosa* and older leaves to *Podosphaera* spp. Early symptoms on foliage include a fine, netlike growth. As infections progress, large numbers of conidia produced on the leaf surface result in a white, mealy appearance (Plate 12). Chlorosis of severely infected leaves may be followed by severe necrosis. The leaves eventually roll upward, pucker, blister, and abscise. Infected shoots are commonly stunted and may be covered with a white, feltlike growth. As colonies of *P. tridactyla* and *P. clandestina* age, numerous brown, spherical cleistothecia are formed, but cleistothecia of *S. pannosa* on stone fruit are rare or absent.

Except for cherry, fruit are infected by *S. pannosa* when young, but they become resistant with maturity. Infected fruit usually are deformed, with slightly depressed or raised areas with mycelium typical of powdery mildews (Plate 13). On older fruit, the epidermis of infected areas may become necrotic.

S. pannosa may infect young fruits in early spring. Fruit lesions may become necrotic and cause surface cracking. Infected areas turn brown or remain green (Plate 14). Mature fruit are resistant to infection by *S. pannosa,* but an unidentified mildew occurring on the apricot cultivar Katy infects young and mature fruit. Infections by *S. pannosa* of newly emergent leaves are noticeable after leaves expand, and the infected epidermal tissue under the mycelium becomes red.

Foliar infections by *P. tridactyla* occur later and are characterized by white mycelia on both leaf surfaces. Severely infected leaves may become distorted and fold longitudinally. When large areas of the leaf blade die, defoliation results.

Cherries—Cherry foliage, shoots, and fruit are susceptible to infection by *P. clandestina.* The initial symptoms, frequently observed several days after bloom, are light, roughly circular lesions on leaf surfaces inside the canopy. As infections progress, sporulating colonies that appear (Plate 12) may remain discrete or coalesce to give the leaves a white, mealy cast. Severely infected leaves frequently blister and pucker. As the colonies age, numerous cleistothecia form. Severely infected shoots are stunted and may appear contorted and blighted.

On ripe fruit, hyphae of the fungus appear in roughly circular, slightly depressed areas on the fruit surface, which may be small and restricted or cover the entire fruit. Young fruit are susceptible to infection. Susceptibility decreases when soluble solids (Brix) exceed 12–13%. In Washington, where the disease is particularly severe, rain showers during fruit development trigger outbreaks of the disease on fruit and increase disease severity on foliage. All sweet cherry cultivars, especially Bing, are susceptible to powdery mildew. Losses on tart cherry are most serious on cultivar Montmorency. Severe petiole infections may interfere with mechanical harvesting.

Peaches and Nectarines—Leaves, young shoots, and fruits are susceptible to infection by *S. pannosa.* Infected leaves may become covered with white, mealy mycelium and may curl and become stunted. On older leaves, the mycelium is white and patchy. Fruit are susceptible from the early stages of growth until pit-hardening. White, circular spots, the first symptoms of fruit infection, may enlarge, coalesce, and cover the entire fruit (Plate 13). Young fruit may be deformed with slightly depressed or elevated areas. On older peach fruit, infected areas may be scabby and necrotic, but on nectarines they remain green. Mycelium usually is absent at this stage. Symptoms on fruit are similar to rusty spot except for the color of the mycelium and lesions.

Causal Organisms

The identification of powdery mildew on stone fruit is difficult because more than one species can occur on a given host. Identification is simplified if ascocarps are present. Cleistothecia of *P. tridactyla* and *P. clandestina* are relatively common; those of *S. pannosa* on stone fruit are rare or nonexistent. The anamorph stages of *S. pannosa* and *P. tridactyla* differ in shape and dimensions of the conidiophore and position of the basal septum. Bases of *S. pannosa* conidiophores are frequently swollen, with the septum 7.5–30 µm from the branching point of the mycelium, whereas bases of *P. tridactyla* conidiophores are generally 0.5 µm from the branching point. Species currently recognized on stone fruits include *P. clandestina* (syn. *P. oxycanthae* (DC.) de Bary), *P. tridactyla,* and *S. pannosa.*

Podosphaera clandestina (Wallr.:Fr.) Lév.: Conidia are produced in chains, which are elliptical to cylindrical, contain fibrosin bodies, and measure 19.9–32.9 × 9.8–21.3 µm (mean 28.6 × 15.0 µm). The fungus is heterothallic. Cleistothecia are numerous, globose to subglobose, occasionally globose-depressed, and black or brown-black when mature; they measure 69.7–110.7 µm (mean 84.3 µm) in diameter. Appendages are infasciculate, five to 20 in number (mean 12.5), 75–282.9 µm long (mean 174.1 µm), one to 3.4 times the diameter of the cleistothecium, and equatorially attached or attached on the upper half of the fruiting body. Appendages are three to four times dichotomously branched, with the primary branches compressed; ultimate branches are swollen and knob-shaped. Appendages are brown-pigmented, at least in the lower half and occasionally up to two-thirds of the appendage length. Asci are ovate and measure 61.2–123 × 35.7–79.1 µm (mean 85.0 × 61.2 µm); ascospores are subreniform to elliptical and measure 20.0–35.7 × 11.9–18.7 µm (mean 29.5 × 16.7µm).

Podosphaera tridactyla (Wallr.) de Bary: Mycelium is white and evanescent to persistent. Conidia (20–32 × 13–18 µm) are ellipsoid to barrel-shaped and are borne on erect conidiophores. Cleistothecia (70–105 µm in diameter) are scattered to gregarious and brown to brown-black. Appendages (one to eight) are inserted in the upper half of the fruiting body, fasciculate or subfasciculate, and unequal in length on the same ascocarp (Fig. 2). Appendages vary in length and are generally one to six times as long as the ascocarp. Appendages branch dichotomously three to five times; primary and occasionally secondary branches elongate. Asci (50–90 × 50–80 µm) are subglobose to ellipsoid-ovoid and lack a distinct stipe. Ascospores (16–30 × 9–20 µm), generally six to eight, are hyaline or cream-colored and ellipsoid-ovoid to subglobose.

Sphaerotheca pannosa (Wallr.:Fr.) Lév.: Mycelium is white; hyphae are thin, 3–7 µm in diameter. Conidia (20–33 × 12–19 µm) are produced in chains from erect conidiophores. Cleistothecia (70–115 µm in diameter) are brown to brown-black, gregarious, and immersed in the mycelial layer. Cells of cleistothecia are irregular-polygonal to rounded and 8–25 µm in diameter. Cleistothecial appendages (generally few) are inserted in the lower half of the fruiting body and are often very short or 0.5–2 times the diameter of the fruiting body. Appendages are septate, hyaline to yellow-brown, and 3–8 µm wide. Asci (70–100 × 50–80 µm) are ellipsoid-ovoid and contain four to eight ellipsoid to cylindrical ascospores measuring 16–28 × 9–18 µm.

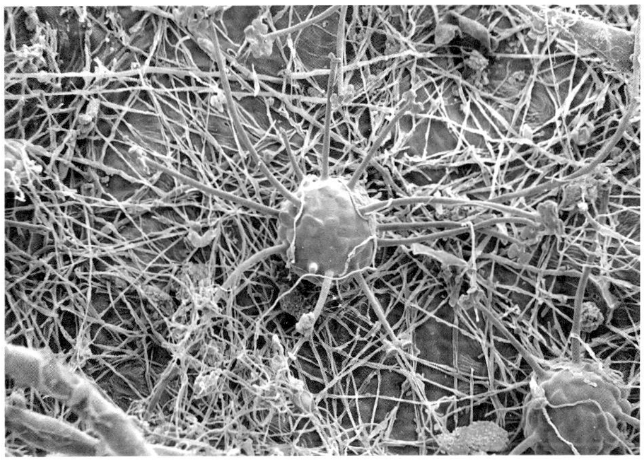

Fig. 2. Cleistothecia of *Podosphaera tridactyla.* Note appendages with dichotomous branching. (Courtesy J. E. Adaskaveg)

Disease Cycles and Epidemiology

Apricot and Plum—There is little detailed epidemiological information on the mildews of apricot and plum. Early-season foliar and fruit infections are caused by *S. pannosa.* Because primary inoculum originates on infected roses or peach near apricot and plum blocks, perennation is thought to occur as mycelium in infected buds of rose or peach. Conidia are wind-dispersed to susceptible tissue. Because *S. pannosa* does not sporulate in profusion on infected leaves and fruit of apricot, and leaf infection on plum is rare to absent, secondary infection is minimal. On plum, primary infections occur on fruit inside the canopy or cluster, but in California, on fruit of apricot cultivar Katy, sporulation continues to initiate new infections until fruit maturity. The species responsible has not been determined. After harvest, *P. tridactyla* can be found on the upper surface of leaves in the inner canopy late in the growing season. Cleistothecia are formed in abundance on both apricot and plum late in the growing season. Because "flag shoots" (similar to those of apple infected by *P. leucotricha*) are absent in spring, perhaps perennation occurs as cleistothecia.

Cherries—*P. clandestina* overwinters as cleistothecia on the orchard floor, trapped in detritus in tree crotches or in bark fissures. Ascospore release requires free water and can occur during wetness durations as short as 1 hr. The optimum temperature for ascospore release is 15°C. Primary mildew colonies are encountered first on leaves of trunk sprouts, those originating from and positioned close to the main scaffold branches, or on leaves positioned immediately above tree crotches. Secondary foliar infections progress throughout the fruit development period and after harvest. All terminal shoots may become infected several weeks after harvest. Conidia germinate best on green fruit and less well as fruit sugar concentration increases up to 15%, beyond which almost none occurs. The optimum temperature for germination on foliage is 20°C. Fruit infections appear if rain occurs near harvest, but it is unclear whether the moisture promotes fruit infection or profuse sporulation of preexisting colonies. In Washington, conidia are present in the orchard air from early May until early or midautumn and are most numerous immediately after harvest in late June. Conidial dispersal is diurnal; concentrations peak in afternoon and are positively correlated with temperature and wind speed and negatively correlated with relative humidity and leaf wetness (Fig. 3).

Peaches—*S. pannosa* overwinters as mycelium in inner bud scales of peach or in dormant infected rose buds. Leaves become infected as they emerge from the buds. Secondary infection of foliage occurs throughout the growing season. Conidia are wind-dispersed. Germination of conidia on foliage occurs from 2 to 37°C and is optimal at 21°C. Foliage is most susceptible to infection at night, when humidity is high and

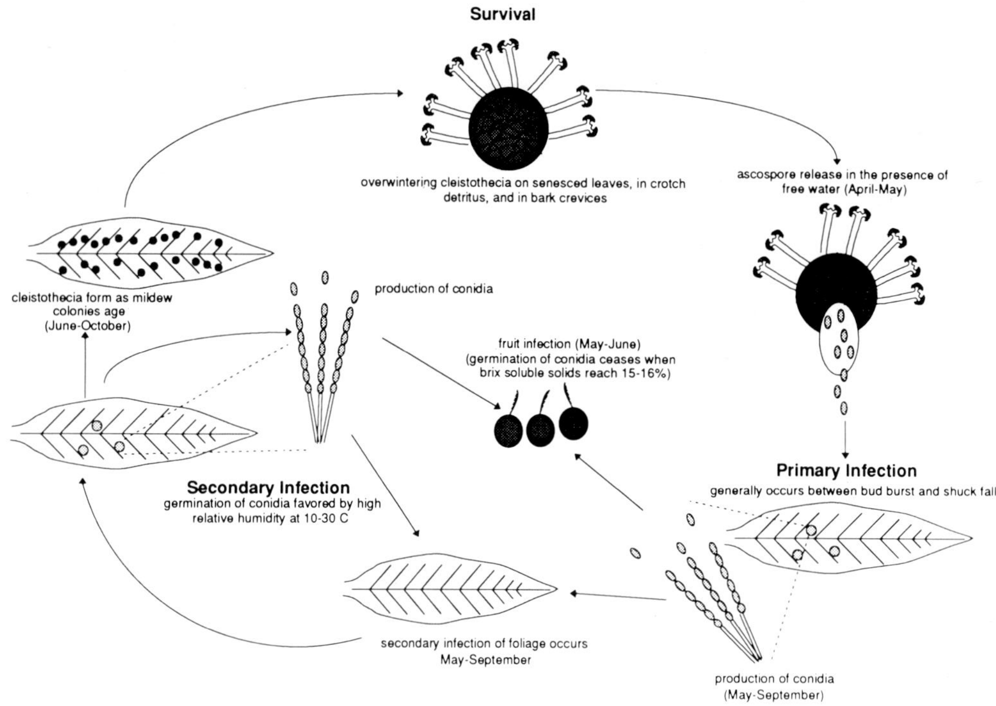

Fig. 3. Life cycle of *Podosphaera clandestina*, cause of sweet cherry powdery mildew.

the osmotic pressure of leaves is low. In Europe, disease outbreaks were less severe when temperatures exceeded 28–30°C or relative humidity was below 70–75%.

Control

Control of mildew on stone fruit depends on the use of fungicides and removal of hosts or the use of resistant cultivars. To control fruit infection, timing of fungicide applications is critical. Removal of susceptible rose bushes near apricot, nectarine, peach, and plum orchards is recommended. Horticultural practices that promote air movement and low humidity should be followed.

Apricots and Plums—On apricots, fruit losses to powdery mildew are most severe on the cultivar Perfection. Cultivars Blenheim, Rival, and Tilton are susceptible. No immunity was identified in extensive testing of European cultivars. Benomyl should be applied first at shuck fall. Sulfur and dinocap may cause phytotoxicity and fruit drop in some areas but have been applied successfully in other areas. In California, removal of rose bushes highly susceptible to mildew in home gardens has effectively controlled mildew without application of protective fungicides.

Plum cultivars Kelsey, Graviota, and Wickson are highly susceptible to *S. pannosa,* but the mildew can be controlled by elimination of mildew-susceptible rose bushes such as Dorothy Perkins. Infection of Black Beaut, Red Beaut, and Ambra plums has been reported in California, and Early Red and Agen are susceptible to *P. tridactyla* in Europe. Protective fungicide sprays are applied starting from the shuck-fall stage of bloom.

Cherries—Applications of fenarimol, myclobutanil, or sulfur control powdery mildew. Flowable sulfur is superior to wettable sulfur. Applications should commence at shuck fall and continue at 7- to 14-day intervals until harvest. In nursery plantings, successful control has been obtained with periodic applications of triadimefon. Water sprouts with initial signs of powdery mildew should be removed from trees between bud burst and full leaf. Cultural practices that promote low orchard humidity should be followed. Black Tartarian, Bing, Chapman, Rainier, and Lapins sweet cherries and Montmorency tart cherries are highly susceptible.

Peaches and Nectarines—In some areas, control is aided by removing infected shoots in winter, removing affected fruitlets when thinning, keeping irrigation to a minimum, and using less susceptible cultivars. Susceptibility to powdery mildew varies considerably and is highest in the nonglandular, pubescent cultivars. In western North America, Flame Crest, Flavor Crest, Red Lady, Elegant Lady, O'Henry, Davidson, Yakima Hale, Peak, and Paloro are susceptible, whereas Angelis, Walton, Johnson, Halford, and Stuart are more resistant. In western North America, most commonly grown nectarine cultivars are susceptible. Powdery mildew can be particularly severe on Red Supreme and Laurie Red.

Sulfur (wettable, flowable, or liquid lime) or myclobutanil is recommended for use on peaches and nectarines. Applications of sulfur or myclobutanil should commence at shuck fall (detachment of dead flower parts from the developing fruit) or early popcorn (petals showing), respectively. Applications of bupirimate or pyrazophos have shown promise for control in Italy.

Selected References

Boeswinkel, H. J. 1979. Differences between the conidial states of *Podosphaera clandestina* and *Sphaerotheca pannosa.* Ann. Phytopathol. 11:525-527.

Grove, G. G. 1991. Influence of temperature and wetness duration on release and germination of ascospores of *Podosphaera clandestina.* Phytopathology 81:1271-1275.

Grove, G. G., and Boal, R. J. 1991. Overwinter survival of *Podosphaera clandestina* in eastern Washington. Phytopathology 81:385-389.

Grove, G. G., and Boal, R. J. 1991. Factors affecting germination of conidia of *Podosphaera clandestina* on leaves and fruit of sweet cherry. Phytopathology 81:1513-1518.

Weinhold, A. R. 1961. Temperature and moisture requirements for germination of conidia of *Sphaerotheca pannosa* from peach. Phytopathology 51:699-703.

Yarwood, C. E. 1939. Powdery mildews of peach and rose. Phytopathology 29:282-284.

(Prepared by G. G. Grove)

Rusty Spot

Rusty spot of peach has been reported in many locations in the United States. The disease can result in significant losses when fruit infection is severe. It has been reported on peach cultivars Rio Oso Gem, Goldeneast, Loring, and Summer Queen, among others. Etiological and epidemiological information on the disease is limited.

Symptoms

The initial symptoms on peach fruit are small, spherical, orange-tan lesions, which gradually enlarge and become indeterminate (Plate 15). Enlargement of lesions parallels fruit enlargement. Fungal mycelium can occasionally be observed in the lesions. Leaves and stems are not affected.

Causal Organism

Podosphaera leucotricha (Ellis & Everh.) E.S. Salmon, described in the *Compendium of Apple and Pear Diseases*, is the suspected cause of rusty spot. Koch's postulates to demonstrate the causal relationship have been fufilled in California, but attempts elsewhere have failed to produce symptoms, or the symptoms were not identical to those of rusty spot in the field. Other evidence supporting *P. leucotricha* as the causal agent includes: 1) control of the disease with sulfur, 2) occurrence on peaches near apple trees infected with *P. leucotricha*, 3) similarity of symptoms to old powdery mildew lesions, 4) hyaline hyphae and conidia that sometimes are observed in rusty spot lesions, and 5) inoculation of fruit with *Sphaerotheca pannosa* or *P. leucotricha* resulting in symptoms typical of peach powdery mildew with the former and typical of rusty spot with the latter.

Epidemiology

Environmental influences upon the development of rusty spot have not been studied carefully. Rusty spot always occurs near apple orchards severely affected with powdery mildew. Usually severity decreases as distance from mildew-affected apple trees increases.

Control

Rusty spot has been controlled when wettable sulfur was applied to peach at shuck split and continued on an eight- to 10-day schedule. Controlling mildew in adjacent apple orchards reduces rusty spot in peach orchards.

Selected References

Daines, R. H., and Trout, J. R. 1977. Incidence of rusty spot of peach as influenced by proximity to apple trees. Plant Dis. Rep. 61:835-836.

Jones, A. L., and Aldwinckle, H. S., eds. 1990. Compendium of Apple and Pear Diseases. American Phytopathological Society, St. Paul, MN.

Manji, B. T. 1972. Apple mildew on peach. (Abstr.) Phytopathology 62:776.

Ries, S. M., and Royse, D. J. 1978. Peach rusty spot epidemiology: Incidence as affected by distance from a powdery-mildew infected apple orchard. Phytopathology 68:896-899.

(Prepared by G. G. Grove)

Green Fruit Rot

Green fruit rot, also known as blossom rot and jacket rot, is a disease of the blossoms and immature fruit of some stone fruits and almonds. Both *Botrytis cinerea* Pers.:Fr. and *Sclerotinia sclerotiorum* (Lib.) de Bary cause green fruit rot; similar symptoms caused by *Monilinia* spp. are discussed in the section on brown rot. Green fruit rot has been reported on apricot in association with *B. cinerea*, *S. sclerotiorum,* or both, in Japan, Italy, the former Czechoslovakia, South Africa, Australia, New Zealand, and the United States. The disease associated with *B. cinerea* has also been noted on cherries in France and California and on plums and prunes in California. Given the cosmopolitan nature of the pathogens, green fruit rot presumably occurs throughout temperate fruit-growing regions of the world whenever weather conditions permit. Epidemics are sporadic but have been reported to reduce apricot yields in California by 70% or more.

Symptoms

B. cinerea attacks all floral parts, and entire clusters of blossoms may be blighted. Infection may spread into the peduncle, and in rare cases it may proceed into the twig, causing a canker. In wet or humid conditions, *Botrytis* can be identified on senescing blossoms by typical gray masses of conidia and by the small sclerotia that are sometimes formed. Conversely, *S. sclerotiorum* does not colonize blossom parts until after they begin to senesce. Its white mycelial mass is evident (Plate 16), but large, black sclerotia are seldom seen on blighted blossoms or infected fruit in situ.

Infection of developing fruit by either *B. cinerea* (Plate 17) or *S. sclerotiorum* (Plate 16) begins at points where the fruit contacts the diseased flower parts. *B. cinerea* may also enter at the point of attachment of an infected style, and where infected petals contact the green fruit. In young apricots, infection causes dark, sunken lesions, which expand rapidly to encompass the entire fruit. The infection is often walled off, resulting in scarring (Plate 16) or deformity in mature fruits. In cherries, green fruit rot results in a smooth brown lesion that is not sunken; the disease is rare unless associated with blossom blight or with doubled fruit in which one twin has aborted.

Causal Organisms

Fungi that cause green fruit rot commonly include *Botrytis cinerea* Pers.:Fr. (teleomorph *Botryotinia fuckeliana* (de Bary) Whetzel), *Monilinia* species (*M. fructicola* (G. Wint.) Honey and *M. laxa* (Aderhold & Ruhland) Honey), and *Sclerotinia sclerotiorum* (Lib.) de Bary. *B. cinerea* forms grayish brown branched conidiophores with a swollen basal cell; oval or elliptical conidia (8–12 × 12–15 μm) are attached to the ampullae by a sterigma. The fungus also generally produces flat sclerotia (shape and size vary considerably) that adhere to the substrate. The apothecial disk of *B. cinerea* is 1–5 mm in diameter. *S. sclerotiorum* produces no conidia, yet produces one to several apothecia (0.5–8 mm in diameter) on hard, irregularly shaped, black sclerotia (0.5–50 mm in diameter). Ascospores (9–12 × 4–6.5 μm) are airborne. Under some conditions, abundant microconidia are formed for both *S. sclerotiorum* and *B. cinerea* that apparently have no function in the disease. In California orchards, apothecia of *S. sclerotiorum* are commonly formed, but those of *B. cinerea* are difficult to find. *M. fructicola* and *M. laxa* are described in the section on brown rot.

Disease Cycle and Epidemiology

Green fruit rot becomes epidemic only following periods of prolonged wet, cool weather. However, the disease has not been extensively studied, and details of its epidemiology must be inferred from work with these organisms in other crops.

Sclerotia of *S. sclerotiorum* that formed on mummified fruit in previous seasons are present in the soil and in debris of the many plants attacked by this fungus. Apothecia are formed in abundance over a broad range of temperatures during the blossoming period in moist, cool soil beneath the ground cover. Ascospores are released from apothecia by a change in relative humidity or by physical disturbance (Plate 18). Infection of floral parts by ascospores requires free moisture for 48–72 hr, but temperature is seldom limiting. The amount of inoculum can be estimated from the relative abundance of apothecia on the orchard floor. There is often more infection in dense fruit clusters, because contiguous fruit are readily invaded.

B. cinerea survives as sclerotia and as mycelia. Sclerotia germinate over a wide range of temperatures to form conidia, and under optimum conditions (15–20°C, high humidity, and light) conidia may be formed in as little as 8 hr. Overwintering mycelia in plant refuse produce conidia as well. Sclerotia may germinate to produce apothecia, but conditions necessary for their formation are not well defined. Release of conidia occurs mainly during the daytime and is mediated by changes in relative humidity. Conidia can be carried long distances on air currents and in water drops, but locally produced inoculum is most important in epidemics. The processes of germination and infection are much more dependent on humidity and wetness than on temperature. In general, the higher the temperature, up to an optimum of 15–20°C, the shorter the wetness period required. Precipitation is not required if relative humidity is above 98%. Tissue exudates and pollen enhance germination, and the presence of injured or dying tissue such as floral parts is necessary in the infection process. The number of fallen blossoms with sporulation can provide a guide to the *Botrytis* spore population. Infections established in green fruit may become latent, to be manifested later when the fruit ripen. Sclerotia are seldom produced in *Botrytis*-infected green fruit unless those fruit fall to the ground.

Control

Control of green fruit rot should be contemplated only when episodes of high rainfall coincide with the bloom and shuck stages. Protective fungicides should be applied during full bloom. Benzimidazole fungicides are effective but should be alternated with nonbenzimidazoles such as captan, iprodione, or chlorothalonil to prevent the development of benzimidazole-resistant strains of *Botrytis*. Fungicide-resistant strains of *S. sclerotiorum* have not been detected. Clean cultivation of the orchard floor may help reduce the amount of inoculum available for infection.

Selected References

Abawi, G. S., and Grogan, R. G. 1979. Epidemiology of diseases caused by *Sclerotinia* species. Phytopathology 69:899-904.

Coley-Smith, J. R., Verhoeff, K., and Jarvis, W. R., eds. 1980. The Biology of *Botrytis*. Academic Press, New York. 318 pp.

Ogawa, J. M., and English, H. 1960. Blossom blight and green fruit rot of almond, apricot and plum caused by *Botrytis cinerea*. Plant Dis. Rep. 44:265-268.

Ogawa, J. M., and English, H. 1991. Diseases of Temperate Zone Tree Fruit and Nut Crops. Publ. 3345. University of California, Division of Agricultural and Natural Resources. 461 pp.

Shotwell, K. M., and Ogawa, J. M. 1984. Botrytis blossom blight and fruit rot of sweet cherry (*Prunus avium*) and their control using fungicides. (Abstr.) Phytopathology 74:1141.

Smith, R. E. 1931. The life history of *Sclerotinia sclerotiorum* with reference to the green rot of apricots. Phytopathology 31:407-423.

Yarwood, C. E. 1948. Apricot jacket rot. (Abstr.) Phytopathology 38:919-920.

(Prepared by K. M. Flint)

Rhizopus Rot

Rhizopus rot is a major postharvest disease on ripe fruit throughout the stone fruit production areas of the world. Losses of up to 50% during the marketing process have been reported.

Symptoms

Rhizopus rot on packaged fruit is characterized by a mass of white mycelium, with long mycelial stolons extending to and infecting adjacent fruit (Plate 19). Black sporangia usually occur in abundance near the edge of the container. Decaying fruit on the orchard floor are covered with black sporangia, but in areas with dry summers the sporangiophores and mycelia are inconspicuous because of low relative humidity. At times, injured fruit on the tree are infected with *Rhizopus* spp., but economic loss in the field is small. The infected tissue is soft and watery. Fungal enzymes in juice leaking from decayed tissue cause brown streaks on the epidermis of adjacent healthy fruit.

Rhizopus rot symptoms can be confused with those caused by *Gilbertella persicaria* (E.D. Eddy) Hesseltine (Plate 19) and *Mucor* spp. Stolons are absent with *G. persicaria,* and mycelia and sporangiophores are erect with *Mucor* spp.

Causal Organism

The disease is most commonly caused by *Rhizopus stolonifer* (Erhenb.:Fr.) Vuill., but *R. arrhizus* A. Fischer and *R. circinans* Tiegh. also may be involved. The optimum temperatures for growth are 27°C for *R. stolonifer* and *R. circinans* and 36°C for *R. arrhizus*. *Rhizopus* spp. are easily cultured on potato-dextrose agar. They grow rapidly with a white, coarse mycelium. Sporulation is almost absent at high relative humidity (Plate 19) but abundant at intermediate relative humidity. Sporangia of *R. stolonifer* are white at first and then turn black. They range from 150 to 350 µm in diameter; the columella is broad, hemispheric, approximately 70–90 µm high and 70–250 µm wide. Sporangiospores are released by disintegration of the sporangial wall; they are unequal, irregular, round or oval, angular, and striated; they are about 7 µm wide and 10–15 µm long when hydrated, but smaller and shriveled when dry. *R. stolonifer* in culture can be confused with *G. persicaria,* but spores of *G. persicaria* have appendages. *R. stolonifer* spores germinate readily at 7.2°C but not at 4.4°C. Zygospores can be produced from matings of compatible strains, but their germination has seldom been observed, and they are not important in infection.

Disease Cycle and Epidemiology

Rhizopus inoculum is derived from growth on organic material in the soil. Uninjured fruit on the tree are not susceptible to infection, but fallen fruit, injured on contact with the orchard floor, are readily infected. The fungus often sporulates heavily on this fruit, and the airborne spores contaminate fruit on the tree. Also, nitidulid beetles feeding on decaying fallen fruit can become contaminated with spores and act as vectors if they feed on, or otherwise injure, mature fruit. Infection of fruit on the tree is relatively uncommon because fresh market fruit are harvested before they are fully ripe. On processing fruit that are harvested ripe, *Rhizopus* decay can develop quickly unless the fruit are protected. Fruit that are harvested mature and ripened at room temperature for drying and freezing are very likely to become infected.

Control

Control for *Rhizopus stolonifer* consists of storing fruit at temperatures below 4.4°C or providing a pre- or postharvest chemical treatment such as dichloran or iprodione plus wax-oil mixture. Dichloran prevents infection from *R. stolonifer* but not other *Rhizopus* spp. Also, special care must be taken to prevent mechanical injury to fruit during harvest, packing, and transit.

Selected References

Ogawa, J. M., and English, H. 1991. Diseases of Temperate Zone Tree Fruit and Nut Crops. Publ. 3345. University of California Division of Agriculture and Natural Resources, Oakland. 461 pp.

Snowden, A. L. 1990. A Color Atlas of Post-Harvest Diseases and Disorders of Fruit and Vegetables. Vol. 1, General Introduction and Fruits. CRC Press, Inc., Boca Raton, FL. 302 pp.

(Prepared by J. M. Ogawa)

Sour Rot

Sour rot is primarily a postharvest disorder of peaches that is characterized initially by a dark brown, wet discoloration of the fruit surface that quickly develops into a wet, mushy decay of the entire fruit. Lesions then become covered with white mycelium and arthrospores or a creamy exudate of mycelium and spores. A sour, vinegar-like odor accompanies the decay. The disease has been reported on peaches in the southern states of the United States. Sour rot caused by *Geotrichum candidum* Link caused major losses of canning peaches shipped from California to the eastern United States in 1993.

Causal Organisms

Two fungi, *Geotrichum candidum* Link and *Monilia implicata* Gilman & E. Abbott, are responsible for sour rot. The mycelium of *G. candidum* is white and prostrate, with few septa. Short, erect conidiophores bear short, cylindrical, truncate conidia in chains. Conidia (5–10 × 4 µm) are single-celled and hyaline. Mycelium of *M. implicata* is white to cream with prostrate conidiophores arising laterally from serial mycelium. Conidia produced in chains are lens-shaped, hyaline, and apiculate and measure 3–4.3 × 1–1.5 µm.

Disease Cycle

Sour rot is associated with poor sanitation in packing sheds. Fruits become contaminated from infested equipment. Chlorination of water and careful sanitation in packing lines are effective for control of sour rot.

Selected References

Burton, C. L., and Wright, W. R. 1969. Sour rot disease of peaches on the market. Plant Dis. Rep. 53:580-582.

Wells, J. M. 1977. Sour rot of peaches caused by *Monilinia implicata* and *Geotrichum candidum*. Phytopathology 67:404-408.

(Prepared by E. I. Zehr)

Miscellaneous Postharvest Fruit Decay

Other postharvest decays are caused by fungi such as *Gilbertella persicaria* (E.D. Eddy) Hesseltine, *Mucor piriformis* E. Fisch., *Mucor* spp., *Penicillium expansum* Link, *Aspergillus* spp., *Cladosporium herbarum* (Pers.:Fr.) Link, *Botryodiplodia theobromae* Pat., *Phomopsis mali* Roberts non (Schulzer & Sacc.) Died., and *Trichothecium roseum* (Pers.:Fr.) Link. These fungi occur naturally in plant residues in the soil. They can be vectored by insects such as the nitidulid beetles and vinegar flies, which carry spores to injured fruit. Thus, one may find these pathogens developing on mechanical and insect injuries on fruit on the tree.

Compared with the major decay pathogens such as *Rhizopus* spp., *Botrytis cinerea* Pers.:Fr., *Monilinia* spp., and *Alternaria alternata* (Fr.:Fr.) Keissl., rots caused by these other fungi are mostly of minor importance, but they occasionally cause significant postharvest losses. Of greatest significance is decay caused by *Mucor piriformis* (Plate 20) and *Penicillium expansum* (Plate 21), because they rot fruit in cold storage, and some species of *Aspergillus*, because they may produce mycotoxins. *A. niger* Tiegh. and *G. persicaria* develop on fruit stored at high temperatures. *A. flavus* Link:Fr., a mycotoxin-producing species, has been isolated but seldom occurs on fresh fruit. Some of these organisms can be important on improperly dried fruit during storage or if insect infestations occur.

These pathogens are less important if fresh fruit are not injured during harvest, packaging, and transportation. Extended storage increases disease severity. The diseases are especially severe on sweet cherry fruit with spurs (aborted fruit) or doubles, on apricots with a cracked stylar-end suture, on peaches and nectarines with split pits, and on prunes mechanically harvested or picked from the orchard floor. On fruit destined for drying, these organisms become established during ripening and storage without refrigeration.

Maintaining clean bins used to transport fruit from field to packinghouse can reduce contamination during processing and handling. Use of chlorinated water in washing solutions reduces the fungal inoculum population on the fruit, but established infections are not killed or suppressed by this treatment. In processing fruit for drying, high temperatures (up to 78°C) and exposure to sulfur dioxide fumigation reduce inoculum viability. Drying fruit to appropriate moisture levels inhibits fungal growth.

Selected References

Ogawa, J. M., and English, H. 1991. Diseases of Temperate Zone Tree Fruit and Nut Crops. Publ. 3345. University of California Division of Agriculture and Natural Resources, Oakland. 461 pp.

Phillips, D. J. 1984. Mycotoxins as a postharvest problem. Pages 50-53 in: Postharvest Pathology of Fruits and Vegetables: Postharvest Losses in Perishable Crops. H. E. Moline, ed. Univ. Calif. Div. Agric. Sci. Bull. 1914.

Snowden, A. L. 1990. A Color Atlas of Post-Harvest Diseases and Disorders of Fruit and Vegetables. Vol. 1, General Introduction and Fruits. CRC Press, Inc., Boca Raton, FL. 302 pp.

(Prepared by J. M. Ogawa)

Alternaria Rot and Red Spot Fruit Blemish

The fungus *Alternaria alternata* (Fr.:Fr.) Keissl. is common in stone fruit production areas. Serious losses on stone fruit are limited largely to fruit during storage; however, Alternaria rot occasionally is a problem in the field on ripe apricots and sweet cherries. A superficial red spot fruit blemish of apricots, caused by *A. alternata*, has been reported in Washington. In California, this condition has been called fog spot, because the red spots are more common when wet weather occurs during fruit set and maturation. A similar disorder on peaches (Plate 22) occurs in some localities in South Carolina and Georgia, where economic losses sometimes result from lost marketability of severely affected fruit.

Symptoms

Alternaria rot is characterized by circular, dry, firm, shallow lesions covered with dark, olive green to black surface mycelial growth. The infected tissue is brown, like that caused by *Monilinia,* but not watery as with *Rhizopus*, and the firm decayed portion can be removed easily from the healthy mesocarp tissue. In storage, Alternaria rot infects mature sweet cherry fruit from infected spurs (aborted fruit); it occurs on the stylar end on ripe and overripe apricots (Plate 22), especially when the blossom style fails to dehisce. In the orchard, Alternaria rot commonly develops near insect feeding injuries or in split pits of fruit. After harvest, established infections can continue to progress in fruit held under cold storage. Alternaria rot may occur on dried fruit if the fruit were infected prior to the drying process. Dried fruit treated with sulfur dioxide are protected from new infections.

The red spot fruit blemish symptoms on apricots appear first as solitary to numerous scattered red dots on the upper fruit surface (Plate 23) about 18 days after shuck fall. Eventually, the spots become light tan to brownish with necrotic centers and red halos. Only the fruit surface and underlying cells are affected.

Causal Organism

Alternaria alternata can be readily identified by its typically muriform (with cross and longitudinal walls) spores that are catenulate (in chains). The conidiophores are simple or branched (up to 50 µm long and 3–6 µm wide); arise singly or in a group; are straight or flexuous, sometimes geniculate, and smooth; and have one or several conidial scars. Conidia (20–63 × 9–17 µm) are formed in long, often branched chains; under magnification they appear golden brown with a smooth or verruculose surface, with up to eight transverse septa and several longitudinal or oblique septa. The beak is short (2–5 µm thick and up to one-third as long as the conidium) compared with that of *A. solani* Sorauer.

This fungus causes fruit rot and was reported as the cause of red spot fruit blemish on apricots in Washington, whereas in California *Alternaria* spp. have been isolated from the "fog spots." The cause of the disorder on peaches in South Carolina and Georgia has not been determined.

A. mali Roberts, which was reported on apples in Japan and in the eastern United States, has not been reported on stone fruit crops.

Disease Cycle and Epidemiology

The causal organism is saprophytic or weakly pathogenic on many plants. In spore trap studies, the concentration of *Alternaria* spores in the air increases during and after rain or foggy conditions. On sweet cherries, the fungus attacks the dead or dying tissue of aborted fruit in the orchard and proceeds into the attached healthy fruit as they mature. In apricots, *Alternaria* frequently infects the stigma after bloom, and when the style remains attached, progresses into ripening fruit. On green immature fruit, germination of spores on the fruit surface causes a superficial blemish, similar to that caused by aphid feeding, but does not appear to progress to decay fruit. *Alternaria* decay is common even under storage temperatures near 0°C and can cause extensive decay of stone fruit stored for periods of two to three weeks.

Control

Deliberate control of *Alternaria* has not been attempted in the field. Diseased fruit and those with mechanical injuries that encourage new infections require hand sorting before packaging. Fruit contamination with *Alternaria* spores is reduced with a rinse in chlorinated water. Effective control has been obtained with postharvest application of iprodione. For red spot fruit blemish on apricots, mancozeb applied in the field before infection reduces the number of red spots. Control of red spot on peaches or fog spot on apricots has not been thoroughly investigated.

Selected References

Larsen, H. J., Covey, R. P., Jr., and Fischer, W. R. 1980. A red spot fruit blemish in apricots. Phytopathology 70:139-142.

Ogawa, J. M., and English, H. 1991. Diseases of Temperate Zone Tree Fruit and Nut Crops. Publ. 3345. University of California Division of Agriculture and Natural Resources, Oakland. 461 pp.

Snowden, A. L. 1990. A Color Atlas of Post-Harvest Diseases and Disorders of Fruit and Vegetables. Vol. 1, General Introduction and Fruits. CRC Press, Inc., Boca Raton, FL. 302 pp.

(Prepared by E. I. Zehr and J. M. Ogawa)

Anthracnose

Anthracnose was first reported on peach fruit in England in 1859. Reports of anthracnose on peaches and other stone fruit followed over the next 50 years from Japan, South Africa, Brazil, and Mediterranean countries. The first report of anthracnose in the United States came from California in 1916 on almond. Significant losses from anthracnose occurred in the late 1940s in the southeastern United States, where peaches typically were graded, packed warm, iced down, and sent directly to market. More recently, outbreaks have occurred sporadically in the southeastern United States and other warm, moist, temperate areas of the world. In 1990, after a heavy rainfall at the end of May, anthracnose on peach and nectarine fruit was detected for the first time in California.

Symptoms

Anthracnose on peach, plum, and sour cherry fruit is characterized by circular, tan to brown necrotic lesions with concentric rings. Orange spore masses often occur in the center. Lesions enlarge rapidly and generally occur in ripe or nearly ripe fruit. On peaches, these lesions are sunken, may reach 4–5 cm in diameter (Plate 24), and are firm to the touch. Lesions may separate cleanly from healthy tissue when pressure is applied. Young lesions may be confused with those of brown rot caused by *Monilinia* species. *Botryosphaeria* species also cause sunken necrotic lesions that separate from healthy tissue, but these lesions are softer to the touch, and spores are not obvious. Another anthracnose of peach in Japan is characterized by a leaf spot, twig canker, and dieback as well as lesions on the fruit. The symptoms are similar to almond anthracnose reported in Israel and California, which is characterized by young fruit infections, leaf spots, and twig dieback.

Causal Organisms

Two species of *Colletotrichum* (*C. acutatum* J.H. Simmonds and *C. gloeosporioides* (Penz.) Penz. & Sacc. in Penz.) cause peach anthracnose. *C. acutatum* is characterized by pink mycelium with orange spore masses on potato-dextrose agar, with conidia averaging 14.8 × 5.2 µm and frequently pointed at one or both ends. No teleomorph has been reported. *C. gloeosporioides* has gray mycelium with orange spore masses on potato-dextrose agar, grows more rapidly than *C. acutatum*, and has slightly longer conidia (16.4 × 5.2 µm) with rounded ends. The teleomorph is *Glomerella cingulata* (Stoneman) Spauld. & H. Schrenk. Perithecia of *G. cingulata* range from 84 to 156 µm in diameter, asci average 65.0 × 11.4 µm, and ascospores measure 18.6 × 5.7 µm.

The causal agent of anthracnose of peach fruit, leaves, and twigs in Japan is reported as *Gloeosporium laeticolor* Berk. The fungus overwinters in the form of mycelia in affected twigs and fruits, on which new crops of conidia are produced the following spring. Conidia (16–23 × 6–9 µm) are ellipsoid or narrowly ovoid, continuous, and hyaline and are produced on slender conidiophores (17–26 × 2–3 µm), which are simple or rarely branched.

Disease Cycle

Glomerella cingulata has a very wide host range including many legumes, herbaceous annuals, and perennials. Blue lupine grown as a ground cover in peach orchards has been associated with increased incidence of peach anthracnose. The fungus is also known to infect apples and grapes. Dissemination of conidia is thought to occur through splashing rain. Infection is generally seen only on ripe or nearly ripe fruit. Disease development is favored by warm (25–30°C), moist weather. Disease cycles for *C. acutatum* and *C. gloeosporioides* probably are similar, but careful comparisons have not been made.

Gloeosporium laeticolor infects young peach fruit through the trichomes or directly through appressoria. Once the young fruit is infected, the fungus grows through the fruit and the peduncle into the phloem of the twig. The infected twig remains alive throughout the winter and dies in the spring. After the twig dies, the fungus forms conidia on the surface of the twig, which serve as inoculum for young fruit. Leaves and older fruit are also susceptible.

Control

Orchard floor and perimeter management to eliminate leguminous hosts and wild *Prunus* species are thought to be useful for control of primary inoculum.

Captan is the only registered fungicide known to be effective for control of anthracnose on peaches. *C. gloeosporioides* is sensitive to benzimidazole fungicides, whereas *C. acutatum* is not. Grower experience indicates that the earlier a captan program is initiated, the better the control. Chlorothalonil has been reported to control *Colletotrichum* spp. on several other crops but is not registered for use on peaches at times when infection of fruits is likely.

Gloeosporium laeticolor can be partially controlled by elimination of infected twigs from orchards.

Selected References

Kitajima, H. 1951. Studies in the dissemination of peach anthracnose II. Ann. Phytopathol. Soc. Jpn. 15:67-71.

Ramsey, G. B., Smith, M. A., and Heiberg, B. C. 1948. Anthracnose of peaches. Phytopathology 41:447-455.

Rittenburg, L. H., and Hendrix, F. F. 1983. Peach fruit rots caused by *Botryosphaeria* spp. and *Glomerella cingulata*. Plant Dis. 67:449-450.

Simmonds, J. H. 1965. A study of the species of *Colletotrichum* causing ripe fruit rots in Queensland. Queensl. J. Agric. Anim. Sci. 22:437-459.

Weimer, J. L., and Dunegan, J. C. 1949. Identity of anthracnose of lupine and peach caused by *Glomerella cingulata*. Plant Dis. Rep. 33:416-418.

(Prepared by B. Bernstein and R. W. Miller)

Sour Pit

Sour pit of peaches may occur as a postharvest disorder first noticeable as abnormal softening of the stem-end cavity of peach fruit. When fruits are cut open, the flesh near the pit is brown to light brown, with slimy appearance and water-soaking of tissues accompanied by a sour or yeasty odor. The yeast *Torulopsis inconspicua* Lodder and Kreger-Van Rij has been reported as the cause of sour pit; it can infect pears, apples, and watermelons as well as peaches. The yeast is transmitted by fruit flies (*Drosophila* sp.).

Selected Reference

Smith, M. A. 1960. Etiology of sour pit of peaches. Appl. Microbiol. 8:256-261.

(Prepared by E. I. Zehr)

Botryosphaeria Fruit Rot

Several *Botryosphaeria* species may cause pre- or postharvest decay of peaches in the southeastern United States. Symptoms are brown to black, sunken necrotic lesions. Diseased tissues are firm and leathery and cling tightly to the mesocarp. In early stages, light or dark bruised areas may appear, and thickened areas dark in color may have a tarlike texture. Symptoms may be confused with anthracnose caused by *Glomerella cingulata* (Stoneman) Spauld. & H. Schrenk.

Causal agents of Botryosphaeria fruit rot are *B. obtusa* (Schwein.) Shoemaker, *B. dothidea* (Moug.:Fr.) Ces. & De Not., and *B. rhodina* (Cooke) Arx. Botryosphaeria fruit rot may be the same disease as Diplodia fruit rot caused by *Diplodia natalensis* Pole-Evans, which is the anamorph of *B. rhodina*. *D. natalensis* was reported to cause postharvest fruit decay in Georgia and South Carolina.

Selected References

Rittenburg, L. H., and Hendrix, F. F., Jr. 1983. Peach fruit rots caused by *Botryosphaeria* spp. and *Glomerella cingulata*. Plant Dis. 67:449-450.

Wright, W. R., and Smith, M. A. 1954. Diplodia fruit rot of peaches. Phytopathology 44:471-472.

(Prepared by E. I. Zehr)

Plum Pockets

Plum pockets is a fungal disease that affects leaves, stems, and fruit of *Prunus* species. It is of little economic importance for commercial plum producers today, but it caused sporadic losses during the latter half of the 1800s and is still found occasionally in Piedmont orchards of the Southeast. The disease is distributed worldwide on various plum types and is caused by several species of *Taphrina*. *Taphrina communis* (Sadebeck) Giesenh. is confined to species of plums (*Prunus angustifolia* and other wild *Prunus* spp.) and was of economic importance in the 1940s on western sand cherry (*P. besseyi*). *T. pruni* Tul. occurs on European plum but is rare in North America.

Symptoms

Symptoms of fruit infection are both more obvious and more prevalent than those on leaves or shoots. Fruit become distorted and enlarged as much as 10 times the normal size (Plate 25), with spongy or hollow centers with or without pits. In modern cultivars, the pockets or bladders can be difficult to detect and may be observed only in cut fruit. Early symptoms are small whitish spots or blisters that enlarge rapidly. Later, they become reddish and display a velvety gray appearance. Upon dehydration, only the exterior shell remains, from which the common names "bladder plums," "mock plums," and "plum pockets" arise. Deformed fruits become brown or black with age and fall from the trees. Infection on leaves and developing shoots results in twisting or curling similar to that produced by *T. deformans* on peach, but symptoms are less conspicuous and not always present. Symptoms on fruit, leaves, and shoots become evident six to eight weeks after budbreak.

Disease Cycle

Disease development of plum pockets is very similar to that of peach leaf curl. The pathogen overwinters on twigs and bud scales as conidia. Infections of leaves, shoots, and fruit occur in wet, cool weather when buds break in the spring. Infections are more prevalent on fruit than on leaves or shoots. The fungal hyphae proliferate very rapidly throughout the developing fruit, producing intense hyperplasia and characteristic bladder fruit. As the plum pockets are formed, the hymenial layer of asci discharges ascospores, and budding ascospores are visible as a velvety gray surface layer on the fruit.

Control

The use of resistant cultivars is the preferred control measure. The susceptibility of hybrids derived from crosses of European and American species is unknown. Many modern cultivars, even though infected, may not produce spores externally, whereas infected wild plums serve as a source of inoculum. Thus, roguing of wild plums near orchard sites to eliminate inoculum sources is very beneficial in control.

The most effective control measure on susceptible cultivars is the application of a single fungicide spray in late autumn or before budbreak in the spring. Effective fungicides include Bordeaux mixture (8–8–100), liquid lime-sulfur, ferbam, chlorothalonil, or other effective fungicides used for peach leaf curl.

Selected References

Anderson, H. W. 1956. Diseases of Fruit Crops. McGraw-Hill, New York. 501 pp.

Buchholtz, W. F., and Nagel, C. M. 1945. Leaf curl and pockets control by a lime-sulfur dormant spray. S. Dak. Agric. Exp. Stn. Bull. 380. 8 pp.

Mix, A. J. 1949. A monograph of the genus *Taphrina*. Kan. Univ. Sci. Bull. 33:1-67.

Swingle, D. B., and Morris, H. E. 1918. Plum pocket and leaf gall of American plums. Mont. Agric. Exp. Stn. Bull. 123:167-188.

(Prepared by K. D. Hickey)

Flyspeck

Flyspeck is a minor problem on plums in Japan and has not been reported on stone fruits in the United States. It is common on apples and pears in temperate areas of the world and occurs on a wide range of other crops including grapes, persimmon, Chinese quince, and pawpaw. The pathogen has been reported on 120 species in 44 families of flowering plants.

Symptoms

Symptoms occur on the surface of mature fruits (Plate 26) and shoots. Numerous shiny, black, superficial, sclerotium-like structures develop in small patches. These patches are usually 1–3 cm in diameter. Conidia and conidiophores of the fungus develop on the sclerotium-like structure during periods of warm, humid weather.

Causal Organism

The causal agent of flyspeck on plum is *Zygophiala jamaicensis* E. Mason. The teleomorph, *Schizothyrium pomi* (Mont. & Fr.) Arx, has not been found in Japan. The sclerotium-like structures produced are similar to pseudothecia described by others, but such structures have not been observed to form ascospores. Conidia (13–20 × 4–6 μm) are two-celled, elliptical to obovate, thin-walled, and often shed in pairs. They are produced on conidiophores consisting of four parts: a modified subhyaline basal cell (16–20 × 4 μm); a conspicuous, tortuous, smooth, thick-walled, dark brown sector (16–18 × 4.5 μm); an angular, subhyaline terminal cell (4–5 × 3–4 μm); and two hyaline conidiogenous cells (6–15 × 4–6 μm) borne on terminal cells marked with few to numerous conidial scars.

Disease Cycle and Epidemiology

The fungus, an ectoparasite, grows on surfaces of plants and probably uses the waxy, cutin layer as its source of external nutrition without penetration into the epicarp tissue. Therefore, the fungus can overwinter as mycelium on the host surface. Primary infection can be initiated on young fruits by airborne conidia produced at, on, and adjacent to the circular sclerotium-like structure. Secondary spread is by chlamydospores and mycelial fragments. Optimum temperature for mycelial growth ranges from 18 to 27°C with relative humidities above 90%.

Control

Since flyspeck is a minor problem on plums, control practices are not used in Japan. Fungicide sprays using ethylenebisdithiocarbamates and benzimidazoles, removal of reservoir hosts, and pruning of trees or thinning of fruit to alter the microclimate may help to mitigate disease development.

Selected References

Nasu, H. 1990. Studies on occurrence and control of flyspeck of grapevine. (In Japanese, with English summary.) Spec. Bull. Okayama Agric. Exp. Stn. 80. 73 pp.

Nasu, H., and Kunoh, H. 1987. Scanning electron microscopy of flyspeck of apple, pear, Japanese persimmon, plum, chinese quince, and pawpaw. Plant Dis. 71:361-364.

Ocamb-Basu, C. M., and Sutton, T. B. 1988. Effects of temperature and relative humidity on germination, growth, and sporulation of *Zygophiala jamaicensis*. Phytopathology 78:100-103.

(Prepared by K. Sawamura)

Diplodina Fruit Rot

Diplodina fruit rot in peaches has been reported twice from Louisiana. The causal fungus infects young twigs, leaves, and fruit. Spots on leaves are irregular, oval to round, and up to 15 mm in diameter; they have alternating brown to black concentric rings with pycnidia scattered over the surface. Sunken lesions on twigs are chocolate-colored and later become dark brown. On fruit, symptoms in early stages of infection are similar to brown rot—small, circular, brown spots that enlarge rapidly. Lesions are soft and watery, but in advanced stages they shrivel in concentric rings. Pycnidia protrude within black stromatic masses sometimes similar in appearance to sporodochia of *Monilinia*. The fungus can be controlled in orchards by sprays of captan.

The causal fungus has been named *Diplodina persicae* Horn & Hawthorne. Pycnidia are dark, ostiolate, spherical, glabrous, erumpent, and approximately 175 μm in diameter. Conidia (26.5 × 6.5 μm) are two-celled (sometimes three-celled), of unequal size, hyaline, and tapered to a blunt point.

Selected References

Horn, N. L. 1952. Diplodina on peach in Louisiana. Plant Dis. Rep. 36:351.

Horn, N. L., and Hawthorne, P. L. 1954. A new fruit rot of peach. Phytopathology 44:134-136.

(Prepared by E. I. Zehr)

Sooty Blotch

The fungus *Gloeodes pomigena* (Schwein.) Colby may infect *Prunus* fruits under conditions of prolonged dampness or high relative humidity, but significant amounts of infection are rare. Colonies of the fungus on fruit surfaces are discrete or diffuse olive green patches. Symptoms, disease cycles, and control are similar to those on apples and pears, which are more susceptible.

Selected Reference

Jones, A. L., and Aldwinckle, H. S., eds. 1990. Compendium of Apple and Pear Diseases. American Phytopathological Society, St. Paul, MN.

(Prepared by E. I. Zehr)

Foliar Diseases

Cherry Leaf Spot

Leaf spot is the most important fungal disease of sour cherry in much of eastern North America and Europe, and also in nursery and orchard plantings of sweet cherry and several cultivated, ornamental, and wild *Prunus* species.

Symptoms

On leaves, lesions first appear as small (3 mm in diameter or less) reddish to purple spots on the upper surface. Lesions on sweet cherry tend to be larger and more nearly circular than those on sour cherry. The spots turn brown and may coalesce. Although affected leaves turn yellow, the area around the lesions may remain green, particularly on sour cherry, giving the leaves a mottled appearance (Plate 27). On some *Prunus* spp., particularly plums, the necrotic circular lesions may drop out, producing shot-hole symptoms. During rainy weather, light pink to white masses appear on the underside of the leaf in the center of the spot. Only a few lesions per leaf can result in leaf chlorosis. Chlorotic, infected leaves abscise, and severe defoliation may result. Infection of fruit pedicels, which can occur when weather conditions are optimum for the disease (Plate 28), results in lesions 3–6 mm long that extend a third to a half of the way around the pedicel. Infection of fruit is rare but may occur in severe epidemics (Plate 28).

Fruit on severely defoliated trees fail to mature normally and are light-colored, low in soluble solids, soft, and watery. They may be smaller or larger than normal, depending on time and severity of defoliation. Early defoliation from leaf spot delays acclimation of wood and flower buds in the fall, hastens deacclimation in the spring, and reduces bud survival and fruit set for at least two seasons. Trees that lose leaves in midsummer are less vigorous and produce fewer blossoms the next year. Trees may die from winter injury if severe cold follows midsummer defoliation.

Causal Organism

Blumeriella jaapii (Rehm) Arx (syn. *Coccomyces hiemalis* Higgins; anamorph *Phloeosporella padi* (Lib.) Arx, syn. *Cylindrosporium padi* (Lib.) P. Karst. ex Sacc.) produces acervuli, which contain conidiophores and conidia, on the lower surface of leaves. Conidia (60–75 × 3 µm) are produced apically on conidiophores (45 µm long × 2.5–3 µm wide at the base) in a sticky mass that is whitish to slightly pink. Conidia are hyaline, two-celled, elongated, and curved or flexuous, with a tapering apex. Apothecia (0.2–0.3 mm in diameter) immersed in host tissues are flat to shallow-convex, globose to obpyriform, and dark brown with subhyaline disks. Asci (50–60 × 12–14 µm) are unitunicate and clavate, with a long, stout, pedicellate base. Ascospores (30–50 × 3.5–4.5 µm) are hyaline, ellipsoid to elongate-fusiform, often slightly curved, and clustered in the upper end of the ascus; they are unicellular or have a median septum. They are separated by filiform paraphyses with a slightly enlarged, hooked or forked apex.

Disease Cycle and Epidemiology

The fungus survives through the winter in leaves on the ground that became infected while on the tree the previous year (Fig. 4). In spring, apothecia are produced in the stroma remaining from the previous season (Plate 29). The optimum temperature for development of asci is 13°C, and the optimum for apothecial development is 16.5°C. Ascospores mature at various times depending on temperature and moisture. Ascospores may be discharged during and shortly after rainfall from late bloom to about six weeks after petal fall. Ascospore discharge is highest at 16–30°C, somewhat less at 12°C, and lowest at 4–8°C. Acervuli also may be produced in the stroma remaining from the previous season, but their role in the epidemiology of leaf spot has not been determined.

The fungus infects through stomata found in lower leaf surfaces, on leaf and fruit stems and, for a brief period, on fruits. Leaves are resistant to infection from budbreak until they unfold, probably because stomata through which the pathogen normally penetrates are not developed. Once unfolded, leaves are susceptible throughout the season, but susceptibility decreases with age. Sweet cherries are less susceptible to leaf spot than sour cherries. The partial resistance of sweet cherry to leaf spot is reflected in longer latent periods, slower lesion expansion, and fewer spores per lesion.

Infection by ascospores and conidia is governed by the duration of wetting from rain relative to temperature. A wet period of only a few hours is required for spore germination and infection when temperatures are favorable. Lesions become visible in 5–15 days, depending on temperature and moisture conditions. Optimum conditions for lesion development are temperatures of 15–20°C accompanied by rainfall or high humidity. As lesions appear, acervuli containing masses of conidia are visible on the undersurface of the leaf. Conidia are carried from leaf to leaf by splashing rain and air currents. Repeated secondary spread and infection by conidia continues until leaf fall in autumn.

Control

Cherry leaf spot is controlled by fungicides such as chlorothalonil, ferbam, dodine, and ergosterol biosynthesis inhibitors. Copper fungicides, particularly Bordeaux mixture, also are effective, but copper rarely is used before harvest because it can adversely affect fruit quality. Applications should begin at petal fall, after the first leaves have unfolded, and at 7- to 10-day intervals to harvest, plus a postharvest application two to three weeks after harvest. Efficiency of fungicide use can be improved by spraying alternate sides on a 7-day schedule

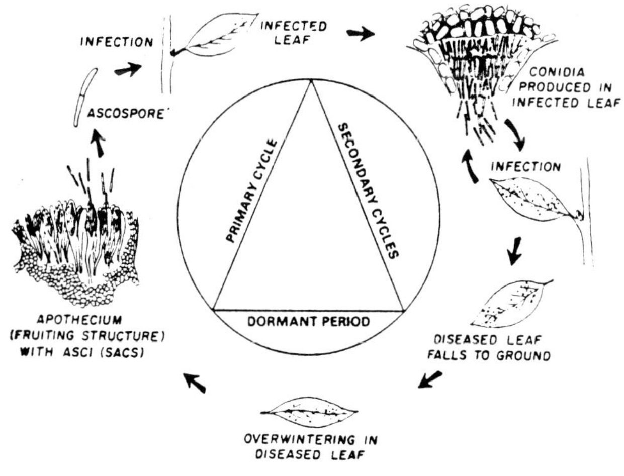

Fig. 4. Disease cycle of cherry leaf spot. (Courtesy A. L. Jones)

rather than spraying each side on a 10-day schedule. Because sweet cherry and European plum are somewhat resistant to leaf spot, the disease on them is easier to control with fungicides than on sour cherry.

All commercially acceptable cultivars of cherry are susceptible to leaf spot, but some are more susceptible than others. Breeding programs exist to develop high-quality cherry cultivars that are resistant to infection by *B. jaapii*. Cultivars with resistance to *B. jaapii* have been developed in some Eastern European countries, but they have not been available for evaluation in other cherry-growing regions.

Selected References

Eisensmith, S. P., and Jones, A. L. 1981. A model for detecting infection periods of *Coccomyces hiemalis* on sour cherry. Phytopathology 71:728-732.

Eisensmith, S. P., Sjulin, T. M., Jones, A. L., and Cress, C. E. 1982. Effects of leaf age and inoculum concentration on infection of sour cherry by *Coccomyces hiemalis*. Phytopathology 72:574-577.

Howell, G. S., and Stackhouse, S. S. 1973. The effect of defoliation time on acclimation and dehardening in tart cherry (*Prunus cerasus* L.). J. Am. Soc. Hortic. Sci. 98:132-136.

Jones, A. L., Ehret, G. R., Garcia, S. M., Kesner, C. D., and Klein, W. M. 1993. Control of cherry leaf spot and powdery mildew on sour cherry with alternate-side applications of fenarimol, myclobutanil, and tebuconazole. Plant Dis. 77:703-706.

Kaszonyi, S. 1966. Life cycle of *Blumeriella jaapii* (Rehm) v. Arx infecting stone-fruits. Acta Phytopathol. Acad. Sci. Hung. 1:93-100.

Keitt, G. W., Blodgett, E. G., Wilson, E. E., and Magie, R. O. 1937. The epidemiology and control of cherry leaf spot. Wis. Agric. Exp. Stn. Res. Bull. 132. 118 pp.

Sjulin, T. M., Jones, A. L., and Andersen, R. L. 1989. Expression of partial resistance to cherry leaf spot in cultivars of sweet, sour, duke, and European ground cherry. Plant Dis. 73:56-61.

(Prepared by A. L. Jones)

Leaf Curl

Peach leaf curl and cherry leaf curl have been reported worldwide from most production areas of peaches, nectarines, and cherries. The disease has been reported on almond in Portugal and apricot in Australia. Economic losses of peach fruit were reported earlier this century from the United States; however, losses of crops in orchards treated with chemicals are negligible. The disease is erratic in the major peach production areas of the United States but occurs every year in the Pacific Northwest and California. It has not been a major problem in the Southeast, but fungicides applied there to control other diseases may effectively suppress the leaf curl pathogen as well.

Symptoms

Yellow to reddish areas appear on young developing leaves in the spring. These areas progressively thicken and pucker, causing the leaf to curl (Plate 30). Puckered areas are crisp and may develop a white coating of spores. Infected leaves abscise prematurely or sometimes remain attached, gradually turning dark brown on severely infected trees. Fruit set on these trees may be severely curtailed. Green shoots also can be infected, becoming thickened and distorted. The rare fruit infections are characterized by irregular, raised, wrinkled, reddish lesions (Plate 31). Cherry leaf curl stimulates excessive branching that results in a witches'-broom symptom (Plate 32).

Causal Organism

Taphrina deformans (Burk.) Tulasne, the cause of peach leaf curl, forms asci on the upper leaf surface. Asci are cylindric-clavate, rounded, or truncate at the apex and have a stalk cell; they vary greatly in size, measuring 17–56 × 7–15 µm. Asci form a compact, whitish hymenium. The eight ascospores, which frequently bud within the ascus, are round, ovate, or elliptic and measure 3–7 µm in diameter. The ascus wall is unitunicate and splits at the apex to release the ascospores. Asci and ascospore characteristics and dimensions are similar for *T. cerasi* (Fuckel) Sadebeck, the cause of cherry leaf curl. Vegetative hyphae are more threadlike than fruiting hyphae and have longer cells. Fruiting hyphae are composed of irregular short cells, mostly located just beneath the cuticle of the upper leaf surface. The mycelium in infected shoots is largely vegetative, and it is intercellular in all host tissue.

Some researchers believe there are varieties or physiologic races of *Taphrina* because symptoms typical of peach leaf curl have not been observed on almonds or apricots in the United States.

Disease Cycle and Epidemiology

Ascospores produced on curled leaves in the spring are forcibly discharged. When ascospores germinate, bud-conidia (blastospores) form, and they continue to bud and eventually cover twigs and buds of the tree. Bud-conidia can reproduce on shoot tips and keep pace with growth of the host all season. Both ascospores and bud-conidia can survive for several months under hot, dry conditions.

Young leaves are infected by bud-conidia, but the time of infection varies among fruit-growing regions, depending on tree development. Infection occurs at budbreak in regions where early bud development occurs quickly. In the Pacific Northwest, infection can occur between bud scales when buds swell slowly during winter months.

Infection can occur on either leaf surface when bud-conidia contact undifferentiated host tissue. Then the fungus changes from a yeastlike saprophyte to a mycelial parasitic phase. Host cells fail to differentiate and enlarge, lose chlorophyll, and may accumulate red pigments in large vacuoles. Only cells in contact with mycelium are affected. Asci develop on the upper surface of hypertrophied tissues.

Periods of cool, wet weather during early bud development favor leaf curl development. The optimum temperature for fungal growth is 20°C, the minimum is 8.9°C, and the maximum is between 26 and 30°C. Budding of bud-conidia occurs at or above 95% relative humidity.

Control

Peach leaf curl is controlled primarily by fungicides and resistant cultivars. Sanitation and cultural practices are ineffective. Fungicides generally are applied before early-season bud development except in California, where sprays are applied shortly after leaf fall to control both leaf curl and shot-hole diseases. No peach or nectarine cultivars are immune to leaf curl, but susceptibility varies; only a few cultivars are resistant.

Selected References

Mix, A. J. 1949. A Monograph of the Genus *Taphrina*. Univ. Kan. Sci. Bull. 33(1). 167 pp.

Ogawa, J. M., and English, H. 1991. Diseases of Temperate Zone Tree Fruit and Nut Crops. Publ. 3345. University of California, Division of Agriculture and Natural Resources, Oakland. 461 pp.

Pscheidt, J. W., and Koepsell, P. A. 1992. Pacific Northwest Plant Disease Control Handbook. Oregon State University Extension Service, Corvallis, OR.

(Prepared by J. W. Pscheidt)

Rust

Rust of stone fruit crops is of worldwide distribution. Its highly variable occurrence is regulated by temperature and rainfall during the growing season. Leaf infections are the usual manifestation. Fruit infections in peach have been reported but seem to be rare. In Brazil on peaches and in France on prunes, complete defoliation can occur unless trees are treated with fungicide. Economic losses are well documented on prunes in Australia and France but not in the United States. In California, rust occurs on prune every year. In some years, early infection results in heavy defoliation that is hastened by mechanical harvesting. The extent and onset of leaf drop is related to the time of initial infection. In southern Georgia (United States), peach leaf rust usually appears in late August or September and does not cause significant early defoliation. However, rust on almond caused early defoliation in several northern California counties in 1983 and 1991.

Leaf rust does not develop until about two months after flowering and does not affect fruit set during the year of infection. In Australia, severe defoliation before harvest reduced yield by causing abscission of immature prune fruit, lowered sugar content, and caused poor drying ratio and small dried fruit. Severe rust epidemics can cause a dry yield loss of 33%. Disease-induced defoliation may reduce tree vigor or productivity in subsequent seasons.

Symptoms

Leaf symptoms are similar on all stone fruit species. Symptoms begin to develop as pale yellowish green spots on both leaf surfaces. The spots are somewhat angular in shape and become bright yellow. Those on the lower leaf surface develop typical orange-brown sori bearing urediniospores (Plate 33). Black teliospores may develop in late season to varied extents among the orange-brown urediniospores, depending on host species and season.

Twig infections are inconspicuous on all host species but are more frequent and more easily found on peach than on prune. Initially, infected twigs show nothing more than slight water-soaking and swelling. Eventually, urediniospores rupture the bark. As twigs continue to grow or resume spring growth, the infected areas split open lengthwise, exposing more urediniospores. Twig lesions are superficial and in themselves have no damaging effect, but they can be an important source of overwintering inoculum.

Fruit infections have been described for peach as water-soaked green spots that become sunken as fruit growth continues. Lesion margins become deep yellow. The yellow-bordered greenish spots stand out vividly against the normal red background color at maturity. Lesion texture becomes leathery. Urediniospores may or may not be present in fruit lesions.

Causal Organisms

Two fungi cause rust on *Prunus* species. *Tranzschelia discolor* (Fuckel) Tranzschel & Litv. has worldwide distribution and is most commonly associated with cultivated stone fruit. *T. pruni-spinosae* (Pers.:Pers.) Dietel is found in Europe and elsewhere, usually on noncultivated *Prunus* species.

The two species of fungi are separated by differences in teliospore morphology. Teliospores of *T. pruni-spinosae* are uniformly coarsely verrucose over both cells, which are the same size, shape, and color and have walls of uniform thickness. The two cells in teliospores of *T. discolor* are not of similar shape or color, and verrucose markings do not cover the entire outer wall. The apical cell is coarsely verrucose and globoid and frequently shows wall thickening at the apex. The basal cell is oblong or ovate-oblong, generally tapered toward the base, and lighter in color than the apical cell. Verrucose markings are reduced or absent on the basal cell. The teliospores are similar in size—those of *T. discolor* (15–23 × 26–39 μm) tend to be somewhat smaller than those of *T. pruni-spinosae* (18–27 × 30–39 μm). The urediniospores are reported to be identical. The urediniospores of *T. discolor* vary in size (9–23 × 18–42 μm) in and among collections.

Disease Cycle and Epidemiology

In most cases, *T. discolor* functions as an asexual rust. The aecial stage, produced on *Anemone coronaria* L., is unknown in many locations where stone fruit rust occurs and is considered insignificant when found. The fungus overwinters as mycelia in twigs or as urediniospores on twigs or on non-abscised leaves. The viability of urediniospores on twig surfaces is greatly reduced during the winter months. Twig infections may produce urediniospores in two successive years. Urediniospores germinate over a wide temperature range (8–38°C; 13–26°C is optimum); thus, the availability of viable spores and moisture, rather than temperature, is considered the major factor limiting infection. The duration of wet periods required for infection at various temperatures is unknown, although an 18-hr wet period at 20°C is more than adequate for heavy leaf infection. The incubation period for twig and fruit infections is not known, but for leaves it is seven to 10 days.

Urediniospores are released in dry wind. Increases in wind velocity result in increased aerial spore concentration. The onset of rainfall or overhead sprinkler irrigation can cause a significant increase in airborne urediniospores. The effects of rainfall on increasing spore release may be purely mechanical, but the exact mechanism is uncertain. Signs of rust are first seen on leaves of young replants. Secondary spread occurs, and at times all the leaves become infected, causing severe defoliation. Premature defoliation can result in emergence of new leaves and blossoms. Infection on the newly developed leaves that remain on the tree during the winter months has been observed.

Control

T. discolor exists in a series of definable formae speciales. The names proposed were *T. discolor* f. sp. *dulcis* for strains that attack almond, *T. discolor* f. sp. *persicae* for strains on peach, and *T. discolor* f. sp. *domesticae* for strains on prune. Peach or almond trees adjacent to severely rusted prune trees are often free of disease. The degree of host specialization worldwide remains unclear, however, and only partial host specificity was found in Australia. There is general agreement that a practical degree of host specialization does exist and may be useful when targeting spray programs to specific species in mixed plantings. Scouting programs might be developed for stone fruit based on observing rust on its most common host species in an area.

Rust can be controlled with preventive fungicide sprays. Treatments one, two, and three months before harvest are applied when early-season disease outbreaks are common. In areas with less frequent or late-season disease problems, fungicides after harvest may be beneficial. Management of rust on French prunes with appropriate timing of applications was possible in Australia with the use of a model that correlates air temperature, rainfall, and the presence or absence of free water in tree canopies with disease development. Where rust is common, updated spray programs are available from local extension offices, agricultural advisory services, or government agricultural bureaus.

Selected References

Bolkan, H. A., Ogawa, J. M., Michailides, T. J., and Kable, P. F. 1985. Physiological specialization in *Tranzschelia discolor*. Plant Dis. 69:485-486.

Carter, M. V., Moller, W. J., and Pady, S. M. 1970. Factors affecting uredospore production and dispersal in *Tranzschelia discolor*. Aust. J. Agric. Res. 21:905-914.

Dunegan, J. C. 1938. The rust of stone fruits. Phytopathology 28:411-427.

Goldsworthy, M. C., and Smith, R. E. 1931. Studies on a rust of clingstone peaches in California. Phytopathology 21:133-168.

Kable, P. F., Ellison, P. J., and Bambach, R. M. 1986. Physiologic specialization of *Tranzschelia discolor* in Australia. Plant Dis. 70:202-204.

Kable, P. F., Milligan, M. J., Wilson, P., Slack, J. M., Ellison, P. J., Watson, A., and Hannah, R. S. 1991. A computer-based system for the management of the rust disease of French prunes. Bull. OEPP (Organ. Eur. Mediterr. Prot. Plant.) 21.

(Prepared by P. F. Bertrand)

Red Leaf Spot

Red leaf spot is an important disease of plum and prune in Germany, Hungary, Serbia, Bulgaria, the former Soviet Union, Iraq, Iran, and Turkey. It is not known to occur in North America. Only leaves are affected. Lesions are circular, elliptical, or irregular blood-red spots that later become black, 6–7 mm in diameter, slightly sunken on the upper side, and raised on the lower side, where dark red ostioles of pycnidia may be seen. Partial or complete defoliation may result. Numbers of flower buds, fruit set, and yield are severely reduced. The causal agent is *Polystigma rubrum* (Pers.) DC. *P. ussuriensis* causes a similar disease in eastern Russia.

P. rubrum overwinters on fallen leaves and bark of affected trees. Wetting of leaves for 26 hr or more results in infection at temperatures between 11 and 30°C. Cultivars differ in susceptibility. Zineb has been reported to control red leaf spot.

Selected Reference

Trifonova, V. 1934. Die Rotflecken Krankheit der Pflaumme *Polystigma rubrum* (Pers.) DC. Phytopathol. Z. 7:73-92.

(Prepared by E. I. Zehr)

Frosty Mildew of Peach

Frosty mildew on peach leaves may be found wherever peaches are grown, but the disease is uncommon except in neglected orchards and nurseries. The fungus, *Mycosphaerella pruni-persicae* Deighton, causes white, powdery patches on lower leaf surfaces similar in appearance to powdery mildew. White patches may be discrete or cover the leaf surface. Indefinite yellowish areas on upper leaf surfaces may be discernible; they become reddish in autumn.

Conidia (17–86 × 2.5–7 μm), borne singly on conidiophores, are abstricted and accumulate in masses. They are hyaline, vermicular to clavate, and several septate. Perithecia may form over winter on fallen leaves if the latter are protected from disintegration. Perithecia mature in the spring and release hyaline, slightly curved, two-celled ascospores (cells of unequal size), which measure 12–20 × 2.5–3.5 μm.

Selected Reference

Higgins, B. B., and Wolf, F. A. 1937. Frosty mildew of peach. Phytopathology 27:690-696.

(Prepared by E. I. Zehr)

Gnomoniosis

Gnomoniosis is a leaf disease of apricot and tart cherry that occurs in France, Austria, Slovakia, Czech Republic, Ukraine, Italy, and the Balkan Peninsula. On apricots, the fungus causes leaf scorch, but on cherry the disease is known as *red spot.* Although the disease is common on leaves in this region, severe economic losses are unusual.

Symptoms

The red spots on cherry leaves are typically surrounded by a large yellow halo. Infected leaves turn brown and remain attached to the stem until the following spring. Leaf scorch symptoms sometimes develop superficially on apricot fruit.

Causal Organism

The causal organism, *Apiognomonia erythrostoma* (Pers.) v. Höhnel (syn. *Gnomonia erythrostoma* (Pers.) Auers), is an ascomycete. The anamorph is *Phomopsis stipata* (Lib.) Sutton.

Disease Cycle and Epidemiology

A. erythrostoma overwinters as perithecia, which form in leaves infected during the previous growing season. Ascospores are released in spring, beginning when the first leaves emerge and continuing for two months. The spores germinate in free moisture and infect by direct penetration of the foliar cuticle. Disease severity is influenced by the amount of available inoculum and the frequency and length of rainy periods.

Control

Triazole fungicides applied when leaves appear in the spring may be used to control gnomoniosis. Zineb, mancozeb, dodine, or dithianon also is effective. Three to five treatments may be needed for effective control. Destruction of fallen leaves in autumn helps to reduce the level of inoculum.

Selected References

Isaeva, E. V., and Bondar', I. A. 1970. Bolezn' abrikosa. Zashch. Rast. (Mosc.) 15:43.

Ponti, I., and Laffi, F. 1988. Maculatura rossa del ciliegio. Inf. Agra. 44:77-78.

Vukovits, G., and Wittman, W. 1990. Identifizierung, Biologie und Bekämpfung des Polzes *Apiognomonia erythrostoma* (Pers.) v. Höhnel, Erreger der Blattbräune kei Marillen. Pflanzenschutzberichte 51:78-90.

(Prepared by E. I. Zehr)

Monilia Leaf Blight and Green Fruit Rot

This disease occurs on sweet cherry (*Prunus avium* L.) and mume (*P. mume* Sieb. & Zucc.) in Japan and on cherry in Korea with comparatively slight losses. It is originally an endemic disease of wild cherries and some early-blooming cultivars of ornamental, flowering cherries such as *P. pseudocerasus* Lindl. and *P. kanzakura* Makino. The pathogen also invades young fruits, and a similar disease in California called "green fruit rot" or "jacket rot" involves *Monilinia* spp. and other fungi.

Symptoms

In early spring, infected young leaves develop small, brown spots that rapidly expand along the veins and midribs. White powdery masses of conidia form on the midribs and veins (Plate 34). Within one or two weeks after blooming, young fruits develop a small brown fleck on the stylar end as infection enters the fruit from the stigma. Later, the entire fruit rots, turns brown, and gradually mummifies. Such mummified fruit is a mixture of host tissue and sclerotia of the fungus—a pseudosclerotium or substratal stroma. Mummified fruits abscise from trees by late summer. Mummified fruits with well-developed sclerotia are light brown, with firm but elastic texture.

Causal Organism

Monilia kusanoi P. Henn. (teleomorph *Monilinia kusanoi* (Takahashi) Yamamoto) causes this disease on stone fruits. The fungus is similar to *Monilinia mali* (Takahashi) Whetzel, which causes Monilia leaf blight of apples. Mummified fruit can survive one or two years on the ground and produce one or two apothecia (Plate 35). Apothecia are cup-shaped to almost flat at maturity, fleshy brown, and 7–11 mm in diameter, with stipes of various lengths. Cylindric, clavate ascospores (140–180 × 7.5–10 µm) contain eight ovoid to broadly ellipsoid, hyaline ascospores (12–15 × 5.5–7 µm).

Conidia produced in chains on simple conidiophores are lemon-shaped to broadly ellipsoid or almost globose and hyaline; they measure 9–11 × 8–10 µm, with disjunctors (structures that develop between mature conidia and facilitate spore liberation) that are 4–5 µm long. Microconidia (approximately 3 µm in diameter) are globose and hyaline.

Cultures on potato-dextrose agar are at first almost white and grow very slowly. The optimum temperature is 20°C for growth of mycelium, which turns brownish black and becomes stromatized (substratal stroma) when aged. The optimum temperature for sporulation of macroconidia is 10°C. Sporulation is highly dependent on alternating light and dark cycles.

Disease Cycles and Epidemiology

Apothecia develop in early spring as leaves of host plants begin to unfold and expand. Early-blooming cultivars are generally highly susceptible. Only ascospores can cause leaf blight, but both conidia and ascospores infect young fruit at the stylar end. Conidia produced on infected leaves serve as secondary inocula for stigma infection of late-blooming cultivars. Wild *Prunus*, particularly *P. jamasakura* Sieb. ex Koidz. and *P. sargentii* Rehd., are reservoirs of the pathogen. In Japan, occurrence on stone fruits, mostly cherry and rarely mume, is local and sporadic. Purely white conidial masses and conidia with disjunctors are the most distinctive features in separating the fungus from *Monilinia fructicola*.

Control

Monilia leaf blight and green fruit rot is controlled by a combination of orchard sanitation and chemical applications. Diseased fruits should be removed and buried or burned. Diseased leaves should also be removed to prevent spread of conidia for stigmatic infection. Plowing the soil in early spring to bury the mummified fruits deep in the soil reduces apothecial formation. Chemical sprays with benomyl or thiophanate-methyl, applied two to three times in a 10-day schedule beginning when leaves begin to unfold, control this disease.

Selected References

Harada, Y. 1977. Studies on the Japanese species of *Monilinia* (Sclerotiniaceae). (In Japanese, with English summary.) Bull. Fac. Agric. Hirosaki Univ. 27:30-109.

Kusunoki, M., Zinno, Y., Kobayashi, T., Minai, J., Hayashi, H., Midorikawa, T., and Iwata, Z. 1984. Studies on the Monilia blight of flowering cherries—Life cycle, epidemiology and disease control. (In Japanese, with English summary.) Bull. For. For. Prod. Res. Inst. 328:1-51, plates 1-4.

Takahashi, Y. 1911. On the *Sclerotinia* diseases of rosaceous fruit trees in Japan. (In Japanese.) Miyabe Hakushi Syushoku 25-nen Kinen Shokubutsu Syusetsu, pp. 135-155, plates 13, 14.

Willets, H. J., and Harada, Y. 1984. A review of apothecial production by *Monilinia* fungi in Japan. Mycologia 76:314-325.

(Prepared by Y. Harada)

Leucotelium White Rust

Leucotelium white rust of peach was first described in Japan in 1912. Because of its white telia, the causal fungus was identified as *Puccinia pruni-persicae*, in contrast to *P. pruni-spinosae*, which produces brown telia. In 1935, Tranzschel established the genus *Leucotelium* to accommodate the white rust organism. The causal fungus also attacks mume (*Prunus mume* Sieb. & Zucc.) and apricots (*Prunus armeniaca* L.). The relative susceptibility of peach cultivars is not known.

The disease occurs only on leaves (Plate 36) and causes serious defoliation and decreased vigor of trees when incidence is very high.

Symptoms

Lesions on peach leaves are polygonal or irregularly shaped, pale green to yellow, and approximately 1 mm in diameter. Cinnamon-colored uredia produced on the underside of leaf lesions resemble those of *Tranzschelia discolor* (Fuckel) Tranzschel & Litv., but the color is somewhat lighter. Telia are produced in late autumn separately or mixed with uredia. Telia are white, slightly mucilaginous, and circular. Affected *Semiaquilegia adoxoides* plants are stunted and usually do not flower, or if they do, flowers fail to develop normally. Small, yellowish brown pycnia develop on the underside of the infected leaves. Cup-shaped, white aecia are produced on the undersurface of leaves near pycnia.

Causal Organism

Flask-shaped pycnia of *Leucotelium pruni-persicae* (Hori) Tranzschel develop from a mass of fungus tissue under the leaf epidermis of *S. adoxoides* leaves. Clear, hyaline, viscous liquid exudes from pycnia along with pycnospores. Aecia, subepidermal in origin, are white, cup-shaped structures that contain aeciospores. Aeciospores (15–21 × 11–16 µm) are oval or ellipsoidal and echinulate (Fig. 5). Uredospores (18–29 × 14–20 µm) are light yellowish brown and broad ellipsoidal or subglobose. The cell wall is uniformly thin and echinulate.

Teliospores (32–44 × 12–14 µm), subepidermal in origin, are fusiform or clavate, two-celled, and constricted at the septum. The cell wall is thin, hyaline, and smooth-surfaced. Each cell of the teliospore germinates to form a short, four-celled hypha (promycelium). Each cell of the hypha bears a hyaline, reniform sporidium (7–9 × 6–7 µm) on a short, pointed sterigma at the upper end of the cell (Fig. 6).

The pathogen is heteroecious. *S. adoxoides,* the alternate host, is widely distributed in Japan, the Korean peninsula, and the People's Republic of China. Accordingly, the known distribution of the disease is limited to these localities.

Disease Cycle and Epidemiology

Aeciospores are forcibly discharged into the air immediately after formation and are carried by the wind comparatively long

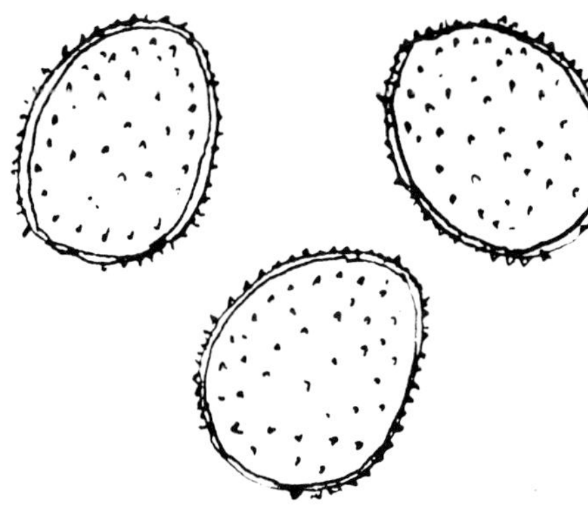

Fig. 5. Oval to ellipsoidal aeciospores of *Leucotelium pruni-persicae* with echinulate markings.

distances. Optimum temperature for infection of young peach tissue is 20°C. Four weeks after infection, light yellow lesions appear first on the upper side of the leaves (usually mid-June). Six to seven weeks after infection, cinnamon-colored uredia that contain uredospores mature on the underside of the leaves. Optimum temperature for germination of urediniospores on a leaf surface is 15–24°C. The germ tube produces an appressorium over a stomatal opening. After an incubation period of 12–15 days, lesions appear; secondary uredia mature six to 10 days later. Telia containing teliospores appear in late October. Spores harvested at this time readily germinate after 2–4 hr at 20°C and produce sporidia after 7–15 hr. Sporidia are wind- or rain-dispersed to leaves of plants during October and December. The fungus invades leaf tissue through the petiole and overwinters in the tuber. Newly developed leaves of *S. adoxoides* are colonized in late February, and pycnia are produced in March, followed by aecia in April. *L. pruni-persicae* can survive in tubers of *S. adoxoides* for several years, producing pycnia and aecia annually.

Control

Little is known about control of white rust. Elimination of the alternate host is recommended for control of the disease. Applications of zineb in early summer may provide effective control.

Selected References

Hiratsuka, N. 1952. On the white rust of peach. (In Japanese, with English summary.) J. Jpn. Bot. 27:229-238.

Hori, S. 1912. A new leaf rust of peach. Phytopathology 2:143-145.

Kitajima, H., and Kishi, K. 1954. The life-cycle of peach white rust fungus, *Leucotelium pruni-persicae* (Hori) Tranz. (In Japanese, with English summary.) Tokai-Kinki Agric.Exp. Stn. (Hortic. Div.) Bull. 2:113-123.

(Prepared by H. Kitajima and K. Sawamura)

Shot-Hole Spot

Shot-hole spot, caused by *Phoma pomorum* Thuem., is an important foliar disease of sweet and tart cherries in Hungary, Georgia, and some parts of the Balkan peninsula. The disease has been reported also from India but does not occur in North America. Plums, prunes, and (in India) apricots also are affected. The fungus causes small, white, brown-edged lesions on leaves, and abscission follows. Cherry fruits can be infected; lesions are spots of concentric zones of brown alternating with carmine red tissue. Sprays of dithianon, thiophanate, or captan have been reported to control shot-hole spot.

Selected Reference

Husz, B. 1939. Hazai adatok a csonthéjas gyümölesfak gombaokozta levélfoltassagauhoz. I. Bull. R. Hung. Hortic. Coll. 5:23-39.

(Prepared by E. I. Zehr)

Target Leaf Spot

Target leaf spot occurs on peach in some parts of Italy and India. The causal agent is *Phyllosticta persicae* Sacc. Although the small circular lesions may be numerous and widespread in some parts of Italy and India, the disease has not been of serious concern. The pathogen occurs on *Prunus* in the United States but is not of economic importance on food crops.

Selected References

Perotti, R. 1942. Note fitopatologiche pergli anni 1939-41. Ann. Fac. Agrar. Univ. Pisa 5:117-135.

Sharma, R. C., and Paul, Y. S. 1986. Target leafspot of peach–A new disease from India. Plant Dis. Res. 1:90-91.

(Prepared by E. I. Zehr)

Zonate Leaf Spot of Stone Fruits

In 1980, zonate leaf spot diseases were first found on apricot and mume in Japan. The range of the disease is restricted to central and northern Honshu (main island Japan), where occasional, local outbreaks occur under prolonged cool, wet weather during the growing season.

Symptoms

Minute, dark green spots that develop on leaves gradually turn gray and enlarge up to 2–30 mm in diameter with concentric rings. The dead or diseased tissues on the spots often fall out, leaving perforations or "shot holes" (Plate 37). On leaves of highly susceptible hosts under humid conditions, water-soaked lesions rapidly develop and coalesce to form large, irregular gray spots on defoliated leaves on the orchard floor. The mycelial masses mature into black, hard sclerotia.

Causal Organism

Cristulariella pruni Harada & Noro, the teleomorph of *Grovesinia pruni* Harada & Noro, causes zonate leaf spot of stone fruits. The fungus is similar to *C. moricola* (Hino) Redhead (syn. *C. pyramidalis* A.M. Waterman & R.P. Marshall, the common zonate leaf spot fungus) but is characterized by its smaller conidia and relatively restricted host range on species of *Prunus*. Other hosts of *C. pruni* include peach and Japanese plum. Sclerotia are tuberoid, black, and 2.5–4.0 mm in diameter. One to several apothecia appear in spring from a single sclerotium, which can survive one or two years in soil on the orchard floor. Apothecia are cup-shaped to almost flat at maturity, fleshy, light brown, and 2–4 mm in diameter, with stipes 2–12 mm long. Asci (140–180 × 9–13 µm) contain eight ellipsoid to ovoid, hyaline ascospores (11–15 × 7–7.5 µm). Sporophores (150–190 × 80–150 µm) are pyramidal, cone-shaped, or broadly ellipsoid and hyaline, consisting of central axis and globose cells. Central axes (approximately 300 × 10 µm) are septate, hyaline, or slightly

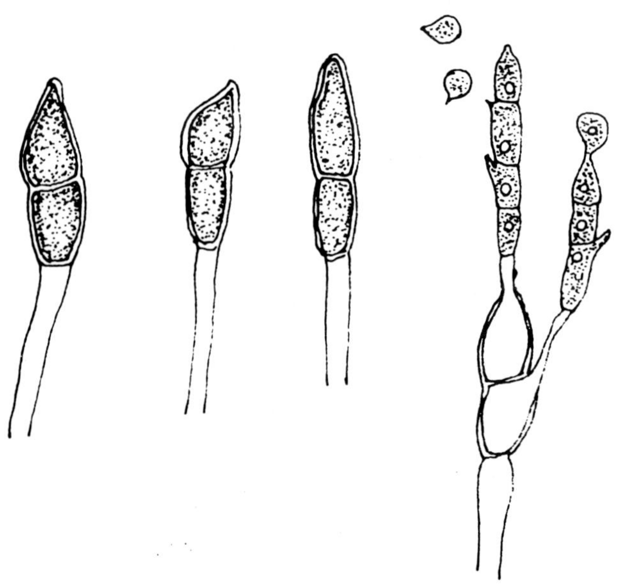

Fig. 6. Teliospores of *Leucotelium pruni-persicae* are two-celled and fusiform or clavate and germinate to produce four-celled hyphae with sporidia on short, pointed sterigmata.

olivaceous. Primary globose cells (approximately 13 µm in diameter) develop pleurogenously from the central axis and reproduce by dichotomous buddings to form secondary globose cells (about 10 µm in diameter), each of which has two small, terminal projections (2–5 µm in diameter). Each globose cell (conidium) may germinate from the terminal projections to produce two hyphae.

Disease Cycle and Epidemiology

The primary inocula for leaf infection are considered to be ascospores from apothecia arising from the sclerotia in late April to early May. The disease is first noticed in June, and symptoms become most obvious in August and September. Conidia produced on lower surfaces of the leaf spots are responsible for secondary infection. Cultivars of apricot are generally more susceptible to the disease than those of mume.

Control

Removal and destruction of diseased leaves are advisable. Chemical sprays with iprodione, thiophanate-methyl, mancozeb, oxine-copper, or Bordeaux mixture, applied twice in midsummer (July and August), provide good control, especially when cool, rainy weather prevails.

Selected References

Harada, Y., and Noro, S. 1988. *Grovesinia pruni*, sp. nov., the teleomorph of a new zonate leaf spot fungus on *Prunus* in Japan. Trans. Mycol. Soc. Jpn. 29:85-92.

Harada, Y., Washio, S., and Matsunaka, K. 1981. Perfect state of a *Cristulariella* causing zonate leaf spot of Japanese apricot. (In Japanese, with English summary.) Bull. Fac. Agric. Hirosaki Univ. 36:12-23.

Noro, S., Sugiyama, S., and Matsunaka, K. 1988. Zonate leaf spot of Japanese apricot caused by *Grovesinia pruni* Harada & Noro. (In Japanese, with English summary.) Bull. Aomori Agric. Exp. Stn. 6:37-71.

(Prepared by Y. Harada)

Septoria Leaf Spot

Septoria leaf spot of plum has been reported in the central United States and Texas. Leaves commonly are infected, but fruit infection has been reported only on the cultivar Bruce. Lesions on leaves are small, circular, necrotic spots that have a distinct tendency to fall out, leaving a pronounced shot-hole appearance. When infections are numerous, leaves may be killed. Lesions on fruits are larger (3–7 mm in diameter), are blue edged with gray, and tend to separate from the surrounding epidermis.

Septoria leaf spot is caused by *Septoria pruni* Ellis. Conidia (25–52 × 2–2.5 µm), found in pale, resinous masses, are three- to six-septate and are produced in subepidermal fruiting structures that are intermediate between acervuli and pycnidia.

This disease seldom causes losses of economic importance, and specific control measures have not been described.

Selected Reference

Lentz, P. L. 1958. Septoria spot of plum fruits. Plant Dis. Rep. 42:1123-1125.

(Prepared by E. I. Zehr)

Cercospora Leaf Spot

Cercospora leaf spot has been reported from many cherry-growing areas, including those in France, Germany, Israel, Italy, and the United States. The disease probably occurs worldwide, although in some countries it is sporadic and not of economic significance. Both sweet and sour cherries are affected by the disease, and many cultivars have been reported to be susceptible to the pathogen, including Chios, Early Black, Windsor, White Esperen, Red Esperen, and Emery. The pathogen also is reported worldwide on other species of *Prunus*, including almonds, blackthorn, plums, and peaches.

Symptoms

Lesions of Cercospora leaf spot disease appear as round necrotic spots that are reddish brown at first; later, as lesions enlarge to 4–5 mm in diameter, the central portion becomes light brown with brownish red edges. Lesions may coalesce to form large necrotic areas (Plate 38) and can develop on both leaf surfaces. Sometimes the necrotic tissue drops out, leaving shot-hole symptoms. The disease causes early defoliation, and severely affected trees may be completely defoliated by the beginning of summer. Premature defoliation may stimulate new growth (e.g., blooming and development of new leaves in autumn), which causes further debilitation of tree vigor.

Causal Organism

Cercospora leaf spot is caused by the fungus *Cercospora circumscissa* Sacc. (syn. *C. cerasella* Sacc.). The teleomorph is reported as *Mycosphaerella cerasella* Aderhold. Mature lesions on leaves are covered with dark brown stromata from which conidiophores and conidia arise. Conidiophores (3–5 × 20–70 µm) are pale brown to brown, sparingly septate, and geniculate. Conidia (2.7–5 × 28–118 µm) are initially hyaline but with maturity become olivaceous, obclavate, base long abconic, straight to mildly curved, and one- to seven-septate.

Disease Cycle and Epidemiology

C. circumscissa overwinters as substomatal stroma in leaf debris on the orchard floor. In the spring, characteristic conidiophores and conidia of the pathogen develop from substomatal stromata of the fungus; conidia are the primary source of inoculum. Conidia from current-season lesions that are disseminated by wind and splashing water produce secondary cycles. High humidity, rain, and dew are favorable conditions for development of the disease. Optimal temperature for disease development is 20–25°C. The importance of ascospores in the epidemiology of the disease has not been reported.

Control

Control of Cercospora leaf spot is based on the use of fungicides. Three to four applications of benomyl, thiabendazole, maneb, or zineb applied at monthly intervals or sometimes more often, beginning at early stages of leaf burst, are effective. Practices that reduce the amount of infected leaf debris in the orchard will significantly decrease the primary source of inoculum in the spring.

Selected References

Jenkins, W. A. 1930. The cherry leaf-spot fungus *Mycosphaerella cerasella* Aderh., its morphology and life history. Phytopathology 20:329-337.

Little, S. 1987. *Cercospora circumscissa*. Descriptions of Pathogenic Fungi and Bacteria, No. 911. Commonwealth Mycological Institute, Kew, Surrey. England.

Sztejnberg, A. 1978. Fungicides controlling the cherry leaf disease caused by the fungus *Cercospora circumscissa* in Israel. (In Hebrew.) Alon Hanotea 32:341-344.

Sztejnberg, A. 1986. Etiology and control of cherry leaf disease in Israel caused by *Cercospora circumscissa*. Plant Dis. 70:349-351.

(Prepared by A. Sztejnberg)

Cankers, Blights, and Wood Rots

Leucostoma Canker

Leucostoma canker, also called perennial canker, Cytospora canker, or Valsa canker, is an important disease of peach, nectarines, and sweet cherries in cooler climates such as in Canada and the northern fruit-growing region of the United States. Leucostoma canker is associated also with the peach tree short life syndrome in the southeastern United States. In California's hot Central Valley, where most of the state's stone fruits are grown, it is of greatest concern in French prune and President plum, both cultivars of *Prunus domestica,* but of relatively minor importance on peach, apricot, and sweet cherry; cultivars of Japanese plum, *P. salicina,* are resistant. In Europe, the disease is important on apricot, peach, and sweet cherry and is part of the disease complex of stone fruits called "apoplexy." Leucostoma canker also occurs in South America and Japan.

The host range includes species within and outside the family Rosaceae, including Sitka mountain ash; blackthorn; black, Japanese, flowering, pin, and sour cherries; chokecherry; Russian olive; apple and pear; cultivated and wild plums; serviceberry; and golden willow. Peach tree cankers caused by *Leucostoma* spp. were first observed in western New York in 1900, a few years later in Missouri, and in southern Ontario in 1912. Leucostoma canker reduces the bearing surface of productive trees and shortens tree longevity.

Symptoms

Symptoms of Leucostoma canker in peach depend on the part of the tree infected. Infections of small twigs appear as sunken, discolored areas, often with alternating zonation lines, usually around winter-killed buds or leaf scars from the previous year's foliage (Plate 39). Nodal infections are easily observed two to four weeks after budbreak. The infected tissues darken with time, and an amber gum may ooze from them unless the twig is killed. Branch cankers that result from such nodal infections will have dead twigs or twig stubs at the center of the canker (Plate 40).

Cankers that form on the main trunk, branch crotches, scaffold limbs, and older branches are the most conspicuous expression of infection. These cankers are typically elliptical and often are accompanied by a copious quantity of amber-colored gum. As the cankers age, the gum becomes dark brown, and the infected bark dries and cracks open, exposing blackened tissue beneath. Leucostoma canker can occur on weakened branches without production of copious gum, especially if winter injury precedes infection.

Beginning in late spring and continuing through summer, the tree grows rapidly and resists further penetration of the fungus into healthy tissues. During this time, the tree may form a callus ring around the canker, but the fungus usually invades this tissue again in late fall or early spring when the tree is dormant and cannot actively resist penetration. The yearly alternation of callus production and canker extension produces a canker with concentric callus rings. When tree defenses are compromised by environmental stresses, callus production may be inhibited and cankers may appear more diffuse.

In California, canker enlargement in President plum and in low-vigor French prune trees is greatest during the summer months. Autumn infection in vigorous French prune trees results in the formation of annual cankers that cease to enlarge the following spring and become surrounded by callus tissue. In President plum and low-vigor French prune, however, the cankers are perennial and result in the girdling and death of affected branches.

Branch or twig infections may produce leaf symptoms during the growing season. Leaves on an infected branch often turn yellow, droop, and may wilt and die. Dead twigs and branches are usually covered with a multitude of pinhead-sized black structures erupting through the dead bark; these are the reproductive structures of the pathogen.

Causal Organisms

The fungi that cause Leucostoma cankers on stone fruits, *Leucostoma cincta* (Fr.:Fr.) Höhn. (syn. *Valsa cincta* (Fr.:Fr.) Fr.) and *L. persoonii* Höhn. (syn. *V. leucostoma* (Pers.:Fr.) Fr.), are ascomycetes in the order Diaporthales, family Valsaceae. The anamorph stages, *Leucocytospora cincta* (Sacc.) Höhn. (syn. *Cytospora cincta* Sacc.) and *L. leucostoma* (Pers.) Höhn.) (syn. *C. leucostoma* Sacc.), respectively, are commonly encountered in the field. The duality in nomenclature that is found frequently in the literature stems from a taxonomic revision of the Diaporthales by von Höhnel in 1917 in which the genus *Valsa* was split into two genera, *Valsa* and *Leucostoma.* The anamorph stages of the two fungi were placed in *Cytospora* and *Leucocytospora*, respectively.

Pycnidial stromata form in cankers and on killed twigs and branches distal to cankers, appearing as quickly as two to three weeks or up to six months after death of the bark. Stromata are black on the surface and gray to grayish brown internally and are underlain by a black layer. Stromal development is first evident as pinhead-sized swellings, the tops of which rupture to expose the disks of the stromata. The disk of *L. persoonii* appears white or frosted, whereas that of *L. cincta* is gray to brownish gray. Mature pycnidia under moist conditions extrude a flesh-colored to orange tendril, or cirrus, containing thousands of conidia. Individual conidia of both species are hyaline and measure 5–10 × 1–2 μm.

Perithecia form much later, often two to three years after pycnidial stromata appear. They may be found within or underneath the pycnidial stromata and may occur on both living and dead branches. Viewed from above, the perithecial stromata of *L. cincta* are round (1.6–2.8 mm in diameter), very prominent, and delimited by a black, circumscribing conceptacle layer (30–80 μm thick), which may be visible through the bark. The ectostromatic disk (0.6–1.1 mm in diameter) is pale gray to tan. Black ostioles of individual perithecia, 10–30 per perithecial stroma, usually are arranged circinately around a central pycnidium, often obscuring the outer portions of the convex disk (Fig. 7). Individual perithecia are 200–350 μm in diameter. Asci (45–80 × 7–12 μm) are clavate and sessile to subsessile and contain eight ascospores (15–30 × 4–8 μm), which are hyaline, aseptate, and allantoid. The perithecial stromata of *L. persoonii* are round (2.0–3.0 mm in diameter), whitish and granular within, and possess a circumscribing conceptacle similar to that of *L. cincta.* Asci are fusoid-clavate (35–45 × 7–8 μm) and sessile to subsessile, and ascospores (10–18 × 2–5 μm) are aseptate, allantoid, and hyaline. Asci and ascospores of *L. cincta* generally are larger than those of *L. persoonii.* The teleomorph stage of *L. persoonii* is fairly common in nature on *Prunus* spp., whereas that of *L. cincta* is relatively rare on *Prunus* spp. and more common on *Malus* spp. For *L. persoonii*, the fungus outcrosses and probably is heterothallic.

The morphological differences between the two species are often difficult to discern; therefore, the species are often differentiated on the basis of their cultural characters on potato-dextrose and Leonian's malt agar, as described by Willison in 1936. Differences include color of mycelium and size and characters of pycnidia. Mycelium is white turning to buff or olive-buff for *L. cincta* and white turning to brown or darker for *L. persoonii*. Pycnidia of *L. cincta* are large (1–3 mm in diameter), white, and felty; they rarely if ever exude cirri, although they are usually full of viable spores. Pycnidia of *L. persoonii* are small (1 mm or less in diameter) and usually dark; they possess a beak and exude cirri when mature. Furthermore, *L. persoonii* grows best at 25–30°C, whereas *L. cincta* grows at 4–30°C (optimum 18–20°C). The ability of *L. persoonii* to grow at 37°C has been used to differentiate the two species in culture. Spore germination requires a carbon source and free water or 100% relative humidity.

Disease Cycle and Epidemiology

Leucostoma pathogens can invade the host only if preceded by a wound or dead tissue. Mechanical and winter injuries and other stresses such as sunburned tissue provide sites for infection. In mature peach orchards where *L. cincta* is the primary pathogen, most new infections appear at the nodes on one-year-old shoots. The pathogen enters the node through leaf scars or dead buds during the dormant season, and the disease may progress to older limbs and scaffold branches. Many new infections also take place at pruning wounds, and the fungus most often isolated at these sites is *L. persoonii*. In general, pruning cuts are most susceptible if they are made early in the dormant season, but many pruning cuts made in the spring can become infected. At temperatures from 14 to 20°C, *L. persoonii* can initiate more rapid canker expansion, but the host response is greater; at 2–8°C, lesion expansion is slower, but with inhibited host defenses the total area colonized is larger. In general, cankers are smaller as the number of days above 10°C increases.

Conidia are the primary inoculum; the potential role of ascospores is uncertain. Conidia are most abundant under the cool, moist conditions of late fall and early spring but are present throughout the year if rainfall is sufficient. Numbers of spores captured are positively correlated with the number of hours that temperatures are between 10 and 15°C, the duration of wetness, and the length of time that relative humidity is above 90%. Conidia are resistant to desiccation when contained within the cirrus; however, after their release by water and subsequent drying, most die within 6 hr. Short-range dispersal of conidia indicates that rain or wind-driven rain may be responsible. Localized splash dispersal also may contribute. Ascospores are discharged forcibly into the air following rain showers at any time of the year when rainfall occurs, but they are most abundant in late spring and early autumn. Spread of *L. persoonii* also may occur via infested pruning tools, birds, and the shot-hole borer (*Scolytus rugulosus*).

In California, the disease on stone fruits is caused entirely by *L. persoonii,* and the most important infection court is sunburn injury on the southwest surface of larger branches on mature trees. The pathogen causes serious damage in young trees only as a secondary invader of cankers caused by *Pseudomonas syringae,* and it is not known to cause infection in nurseries. Shot-hole borers, which commonly are attracted to stressed or injured branches, create entry points for the pathogen, and a low percentage serve as vectors.

Control

Leucostoma canker can be controlled only within the framework of an integrated crop management strategy. Management of cankers is based on preventive measures that minimize winter injury, sunburn, and insect damage, promote optimum plant health, and facilitate rapid wound healing.

Proper site selection for new plantings is essential. Deep, well-drained soil and good air drainage minimize winter injury. New plantings should be reasonably isolated from sources of inoculum such as older, heavily infested orchards and should not be established on the downwind side of older blocks.

Nursery stock should be disease-free and not larger than 11/16 caliper. Trees with small cankers on lateral branches may be planted if at least 10 cm of healthy tissue below the canker is removed. All trees should be examined closely, and trees should be planted immediately after they are received from the nursery to avoid additional stress. The roots and crown of new trees should be dipped in an appropriate insecticide to protect them from the peach tree borer. Newly planted trees should be pruned when budbreak begins and headed back to about 100–115 cm to promote wide-angled branching. Small trees can be pruned to whips, but on larger trees four to six side branches should be pruned to two or three nodes, since trunk buds may not develop. After growth begins, any dead branches should be removed.

Trees must be trained during the first season to avoid narrow crotch angles that result in susceptibility to winter injury, invasion by borers, and proneness to split when bearing a heavy crop. Wire spreaders or wooden spreaders with nails that injure bark should be avoided.

Oriental fruit moths and peach tree borers can seriously damage young trees, and their feeding activity creates infection sites for *Leucostoma* spp. It is also important to avoid twig infections by the brown rot fungus, because they are often invaded and enlarged by *Leucostoma* spp.

Rodent damage should be prevented with wire or plastic guards. Plastic wraparound guards should be removed each summer, because they may delay hardening of the wood in late fall, may harbor boring insects, and interfere with trunk sprays for borer control. Latex paint with thiram also discourages rodent feeding.

To avoid low-temperature injury, excessive or late fertilization with nitrogen and late-season cultivation should be avoided. However, trees with nitrogen-deficient leaves are more susceptible to infection by *Leucostoma* spp. than those that are well fertilized. Nitrogen fertilizer application should be balanced with an adequate supply of potassium. Leaf analysis should be used to determine fertilizer requirements. In clean-cultivation management systems, cultivation should be

Fig. 7. Perithecial stroma of *Leucostoma persoonii.*

stopped and a cover crop sown within three weeks of early fruit drop. Sod management with trickle irrigation, in addition to maintaining tree growth and fruit size, has the added benefit of making trees more resistant to *Leucostoma* spp.

Southwest injury or sunscald caused by warming of the bark by direct sunshine on the south and west exposures of the trunk and scaffold limbs may occur even during relatively mild winters. These sites are commonly infected by *Leucostoma* spp. To avoid southwest injury, trunks and scaffolds should be covered with white latex paint to reduce bark temperatures on sunny winter days. In warmer regions, such as California's Central Valley, a modified pruning system that maintains better leaf canopy over the larger limbs has markedly reduced sunburn injury and Leucostoma canker in bearing President plum trees. With French prune, effective control includes providing good tree nutrition and adequate irrigation, avoiding shallow soils above a "hardpan" and soils with a heavy clay content, correcting potassium deficiency, and reducing the population of the ring nematode *Criconemella xenoplax.*

Small mounds of soil or mulch that drain water away from the tree trunk may prevent direct cold-temperature injury to the crown and physical injury from formation of ice collars. Gravel should not be used to fill depressions around tree collars.

To promote rapid wound healing, pruning should be delayed until the first forecasts of warm, dry weather. Approximately 390 accumulated degree-days (base 0°C) are required for complete wound healing. In general, any practice that promotes tree health encourages more rapid healing. Pruning should be well planned each year so that large cuts, which heal more slowly, will not be needed. When pruning, avoid leaving stubs, which may become infected. The ridge of thickened bark where smaller branches join the larger limb should not be removed, because the most rapid wound healing occurs in this region. In areas where sunburn is not a factor, prune to open the center of trees to light penetration, because shaded branches are weak and susceptible to winter injury and *Leucostoma* infection. All dead and weakened wood should be removed.

Cankers should be removed from the tree and burned, buried, or moved out of the orchard. Cankers on trunks and large limbs can be removed surgically in midsummer, when trees heal most rapidly. Surgery should be performed in dry weather that is expected to persist for at least three days. During surgery, all diseased bark around the canker and about 3–5 cm of healthy surrounding tissue should be removed. Cutting tools should be disinfected between cuts with an alcohol or bleach solution. The resulting wound when finished should have a smooth, slightly rounded margin above and below to favor rapid wound closure. Sites of surgery heal best if left uncovered.

The practice of covering pruning cuts in spring with a thiram-latex paint mixture provides some degree of protection against fungal infection. Benomyl may reduce leaf scar infections by *L. cincta* that accompany defoliation in autumn. No fungicides are registered specifically for control of *Leucostoma* spp.

Selected References

Bertrand, P. F., and English, H. 1976. Release and dispersal of conidia and ascospores of *Valsa leucostoma.* Phytopathology 66:987-991.

Bertrand, P. F., English, H., and Carlson, R. M. 1976. Relation of soil physical and fertility properties to the occurrence of Cytospora canker in French prune orchards. Phytopathology 66:1321-1324.

Biggs, A. R. 1989. Integrated approach to control Leucostoma canker of peach in Ontario. Plant Dis. 73:869-874.

Northover, J. 1976. Protection of peach shoots against species of *Leucostoma* with benomyl and captafol. Phytopathology 66:1125-1128.

Schulz, U., and Schmidle, A. 1983. Zur Epidemiologie der Valsa-Krankheit. Agnew. Bot. 57:99-107.

Tekauz, A., and Patrick, Z. A. 1974. The role of twig infections on the incidence of perennial canker of peach. Phytopathology 64:683-688.

Willison, R. S. 1936. Peach canker investigations. II. Infection studies. Can. J. Res. Sect. C 14:27-44.

(Prepared by A. R. Biggs)

Silver Leaf Disease

Silver leaf (silver blight), a destructive disease of stone fruit trees, is present in most temperate zone production areas, including North America, Europe, Asia, South Africa, Australasia, and South America. It is a major economic disease of stone fruit in France, Australia, New Zealand, Chile, and other countries. Plums, peaches, nectarines, and cherries are most commonly affected, but almonds and apricots are also susceptible. The disease also occurs in fruit tree nurseries, and a wide range of deciduous ornamental and shelter species, such as willow and poplar, may serve as alternative sources of infection.

Symptoms

Leaf silvering is the characteristic symptom on most fruit species and occurs soon after the fungus gains entry into the trees. Fungal toxins carried in the xylem from infected limbs cause the upper epidermis to separate from the palisade layer, generating the silvering effect. Severely affected leaves are silver-gray, curl upward slightly at the edges, and may become necrotic (Plate 41). The silver-gray color contrasts dramatically with the dark green of adjacent healthy foliage. Brown heartwood staining indicates the location of the fungus in the tree. Once a tree (or limb) is killed by the fungus, fruiting bodies (basidiocarps) that appear on the surface of the dead bark are a sign of the fungus. They are typically 2.5 cm across and are often arranged in tiers (Plate 42).

Causal Organism

Chondrostereum purpureum (Pers.:Fr) Pouzar (syn. *Stereum purpureum* (Pers.:Fr)), a basidiomycete in the family Stereaceae, forms a thin, leathery sporophore 2–8 cm wide, which is variably colored from gray to brown and fades with age. The structure is circular, resupinate or profusely imbricate, with a hairy upper surface. The hymenium is smooth and variable in color from fawn to purple. The basidiospores (6–8 × 3–4 μm) are hyaline, oval, and apiculate at one end. A characteristic histological feature is the presence of vesiculose, subhymenial cystidia, 15–30 μm long and 12–25 μm wide, extending above the basidia.

Disease Cycle and Epidemiology

The pathogen is spread by basidiospores that are released into the air from sporocarps during and after rainfall when the relative humidity is above 90%. Spore release occurs at temperatures between 4 and 21°C and at any time of the year, day or night. Spores are shed successively from the same sporocarps for up to two years. Infection follows when the basidiospores land on a fresh pruning cut that exposes xylem tissues. Spores deposited on the wood during moist weather germinate in place or are drawn into the xylem vessels, where they germinate to produce hyphae that spread in the wood. Infection may also spread by natural root grafts in cherry, but this is not common. Diseased alternative hosts such as shelter and ornamental trees are sources of inoculum, as are stumps or stacks of recent prunings from diseased trees.

Pruning cuts are most susceptible during the first week after wounding and are possibly resistant to infection after one month. The fungus is confined to the xylem and does not grow in cambial tissues.

The risk of infection and disease is enhanced if trees are pruned during late winter and early spring, when concentrations of nutrients in sap, such as nitrogen and carbohydrates, are highest. The most vigorous cultivars tend to be the most susceptible. Infection also is favored when trees are forced into rapid growth by heavy pruning, particularly in nurseries when rootstocks are headed back to the scion buds in early spring.

Control

Control is difficult because the pathogen has many hosts, including common shelter and garden species. Widespread infection can result from basidiospore "showers" whenever rain falls in areas where the disease is common. Because it is virtually impossible to fully protect all wounded surfaces, approximately 10% of trees may contract silver leaf each year in areas where the disease is severe.

To reduce the risk of disease, all potential sources of live basidiocarps, such as prunings, shelter belt trimmings, stumps, or logs, should be burned or buried before sporocarps develop. Pruning should be done when little rainfall is expected and when trees are less susceptible (summer to autumn). Immediately after pruning, larger wounds should be treated with a fungicidal wound dressing. Although these have not been effective in the past, newer formulations show promise. Automated methods of applying fungicidal wound dressing with modified pruning shears are available.

In recent years, curative biological treatments have been developed, based mainly on selected strains of *Trichoderma viride* Pers.:Fr. Remission of symptoms has been achieved by inserting dowels impregnated with *Trichoderma* into the trunk below the diseased limbs or by injecting a liquid formulation. Other promising antagonists are *Truncatella* spp. and *Gliocladium* spp. Fungicide solutions for trunk injection also are being developed.

Selected References

Cunningham, G. H. 1922. Silver blight, *Stereum purpureum* Pers., its appearance, cause, and preventative treatment. N.Z. J. Agric. Res. 20:276-283.

Dubos, B., and Ricard, J. L. 1974. Curative treatment of peach trees against silver leaf disease (*Stereum purpureum*) with *Trichoderma viride* preparations. Plant Dis. Rep. 58:147-150.

Dye, M. H. 1972. Silver leaf disease of fruit trees. N.Z. Minist. Agric. Fish. Bull. 104. 20 pp.

Ogawa, J. M., and English, H. 1991. Diseases of Temperate Zone Tree Fruit and Nut Crops. Publ. 3345. University of California Division of Agriculture and Natural Resources, Oakland. 461 pp.

(Prepared by K. G. Tate)

Constriction Canker

Constriction canker of peach, also known as Fusicoccum canker, peach canker, stem canker, constriction disease, and *viruela de la pua,* was devastating during the 1940s in peach orchards in the Middle Atlantic states. Some peach cultivars in New Jersey, such as Golden Jubilee, were so severely affected that by the mid-1950s only a few orchards of these cultivars remained. The disease also occurs on peaches in France, Italy, Japan, Portugal, Bulgaria, Argentina, Brazil, the United Kingdom, Sardinia, and Tunisia. Almond was the first host on which the disease was observed. Dieback of plums from constriction canker has been reported in England.

Symptoms

Leaves and twigs are usually the only tissues affected, but a firm, brown fruit decay also has been observed in New Jersey. On leaves, the pathogen induces large, brown, zonate spots that are circular or irregular in outline. Centers of lesions become sparsely dotted with black pycnidia. The fungus is restricted within the leaf spot during hot weather but grows into the veins as the leaves senesce in the fall.

On twigs, reddish brown to brown elongate lesions (cankers) are centered on an infected bud or node of one-year-old shoots (Plate 43). Lesions first become visible early in the spring, and as they enlarge, they cause girdling and a blighted appearance of shoots. The number of newly blighted shoots may continue to increase into the summer. Lesions sometimes are confused with blossom blight caused by *Monilinia laxa* or *M. fructicola*, but lesions of constriction canker tend to be sunken and centered on a bud or node rather than a blossom, and they often have a zonate pattern that is visible on the surface and in the phloem of infected tissues. A small amount of gum often exudes from lesions of constriction canker, but the profuse gumming associated with blossom blight caused by *Monilinia* is absent.

Causal Organism

The fungus *Phomopsis amygdali* (Del.) Tuset & Portilla (syn. *Fusicoccum amygdali* Del.) causes constriction canker, but there has been confusion in the literature as to the name of the causal agent. Some have used *Phoma persicae* Schulzer & Sacc., which apparently can cause similar twig lesions on peach, as the name for the pathogen; others have used *Phomopsis mali* Roberts non (Schulzer & Sacc.) Died. (teleomorph *Diaporthe perniciosa* Ém. Marchal or *D. eres* Nitschke) or *P. amygdalina* Canonaca, which causes withering of peach shoots. Additional changes in nomenclature may become necessary as characteristics of the fungus are understood.

Pycnidia of *P. amygdali* are usually solitary, eustromatic, globose or applanate, sometimes ellipsoidal, dark brown to black, 180–550 μm in diameter, and ostiolate, with an indistinct or clearly visible papillate ostiole. Conidiophores are hyaline, cylindrical, sparingly branched, and septate at the base, with one or two septa. Alpha conidia (4.9–9.9 × 2.3–3.8 μm) are hyaline, fusiform, straight with slightly acuminate ends, and unicellular. Beta conidia (14.3–23.1 × 1.1–1.9 μm) are rare or not found; when present, they are hyaline, filiform, elongate, straight or more often slightly curved, and unicellular. Conidia exude in cirrhi (Plate 44).

In culture, *P. amygdali* grows well on potato-dextrose agar in a lobulate form. Ability to sporulate is lost after several transfers. Mycelium is hyaline, branched, and septate and often forms chlamydospores in chains.

Disease Cycle and Epidemiology

Twig infection generally is more common and injurious to trees than is leaf infection. It often occurs in autumn near leaf fall when leaf scars are exposed to invasion, but infection also occurs in early spring. Invasion also may occur through buds, bud scale scars, stipule scars, fruit scars, or blossoms, or even directly through young shoots. Summer infections may occur if mechanical injury coincides with moist weather. Development of lesions and spread of infection are interrupted by cold winter temperatures but resume with warm temperatures in the spring. Cankers form at the base of new shoots, and if the twig is girdled, the shoot distal to infection dies. The fungus secretes a toxin (fusicoccin), which stimulates guard cells of stomata to remain open, thereby hastening drying and withering of leaves.

A single disease cycle per season is typical, but at least one secondary cycle may occur if weather conditions favor infection. Pycnidia form in infected tissues, and conidia exude in tendrils (cirri) during moist periods. Conidia exude rapidly at 10°C but slowly at 5°C. Conidia are aerially disseminated after release by rainfall. They germinate on moist surfaces at 5–36°C (optimum 27–29°C). Successful invasion of host tissue is most common at 5–15°C, perhaps because at temperatures near the optimum for germination, openings in plant tissues heal before invasion is accomplished.

Control

Peach cultivars differ in susceptibility to constriction canker, but few are resistant. Red Haven, Golden Jubilee, Rio Oso Gem, Raritan Rose, and Redgold nectarine are most susceptible, whereas J. H. Hale, Dixired, Harken, Sunhigh, and Coronet are less susceptible.

Certain fungicides, notably the benzimidazoles, chlorothalonil, captan, and captafol, help to prevent constriction canker if applied prior to infection. Copper plus lime is effective if applied just before budbreak and may be applied also in several fall applications when the disease is especially troublesome. The widespread use of captan and benzimidazole fungicides to control scab and other peach diseases during the growing season might be a factor in the diminished importance of this disease in the United States.

Careful pruning and destruction of cankered twigs help to eradicate sources of inoculum in peach orchards. Pruning and judicious use of fungicides help to control epidemics of constriction canker.

Selected References

Cohoon, D. F., and Daines, R. H. 1956. Peach canker (*Fusicoccum amygdali*): Times and sites of infection. Plant Dis. Rep. 40:304-308.

Guba, E. F. 1953. Large leaf spot and canker of peach caused by the fungus *Fusicoccum amygdali* Delacr. Plant Dis. Rep. 37:560-564.

Guba, E. F. 1958. Pertinent aspects of the infection cycle and control of the peach canker pathogen, *Fusicoccum amygdali*. Plant Dis. Rep. 42:481-498.

Jailloux, F. 1985. Sensibilité de 27 variétés de pêcher vis-à-vis de *Fusicoccum amygdali* Del. agent du chancre de rameaux. Agronomie 5:515-520.

Montero, J. C., Espósito, S. M., González de las Heras, B., and Marsili, R. C. A. 1987. Evaluación de fungicidas en el control de la viruela de la pua (*Fusicoccum amygdali*) en duraznero. Bol. Téc. Est. Exp. Mercedes 7(4):71-76.

Tuset, J. J., and Portilla, M. T. 1989. Taxonomic status of *Fusicoccum amygdali* and *Phomopsis amygdalina*. Can. J. Bot. 67:1275-1280.

(Prepared by E. I. Zehr)

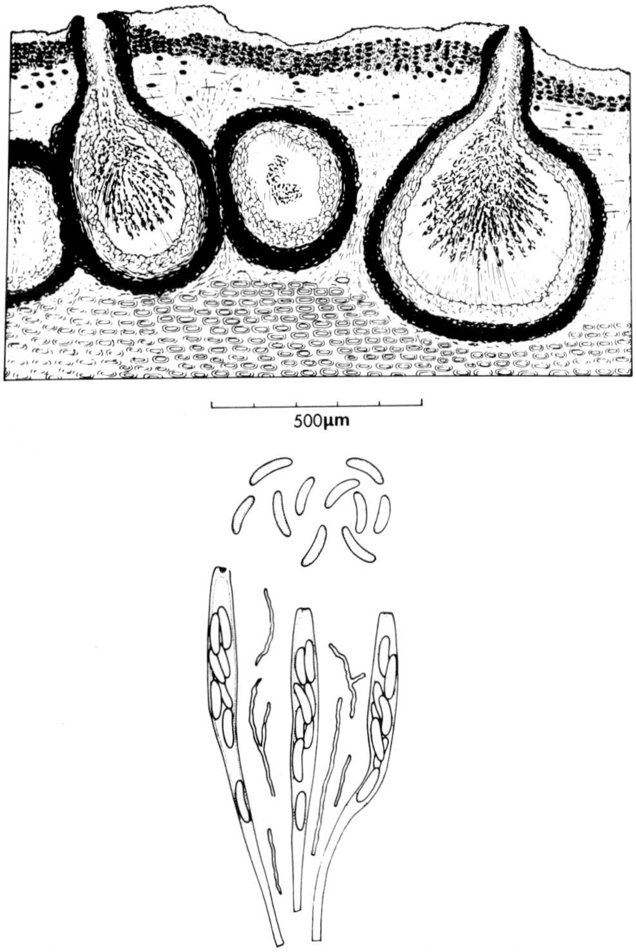

Fig. 8. Perithecia embedded in the stromata (top) and asci and ascospores (bottom) of *Eutypa lata*. (Reprinted, by permission, from Carter, 1991)

Eutypa Dieback

Eutypa dieback, also known as *Eutypiose* in some French-speaking countries and formerly known as "gummosis disease" in Australia, is one of the most destructive diseases of apricot. Sometimes it also affects almond trees and some cultivars of European plum. The disease occurs widely throughout temperate regions of the world and is most abundant where natural rainfall is 60 cm or more.

The causal fungus, an ascomycete formerly known as *Eutypa armeniacae* Hansford & Carter and now designated *Eutypa lata* (Pers.:Fr.) Tul. & C. Tul., has a known host range of nearly 90 species distributed in 28 botanical families, including many natural forest and ornamental trees. In some hosts, the impact of the pathogen is neither severe nor obvious to the casual observer.

Symptoms

Eutypa dieback may be seen in apricot trees of almost any age, but its incidence increases rapidly from year to year as the trees grow. The first visible symptom is a branch canker, most commonly centered on a pruning wound made several years previously. Extending upward and downward from the canker, a zone of discolored sapwood may be found, usually terminating in a light brown and somewhat diffuse margin. The extent of this zone may be revealed by slicing the bark from the sapwood with a sharp blade. The zone of invasion may gradually extend for several years to encompass the entire cross section of the branch. Leaves on the portion of the branch beyond an established canker may wilt suddenly, signaling death of that part of the branch. Leaf wilting occurs typically in mid- to late summer, often when the tree is bearing a heavy load of fruit, which exacerbates moisture stress (Plate 45).

On branches killed by the pathogen, leaves fail to abscise and often remain attached through the following winter. The dead wood assumes a brittle texture and, when bent, snaps easily and suddenly with a short-grained fracture, similar to the break that occurs when a crisp carrot root is bent sharply. Branches that have died from other causes, by contrast, when bent will almost invariably fracture gradually, with a stringy break. If the dead wood remains on the tree or on the ground for two or more years, a heavy black stroma on the surface of the sapwood causes bark to slough off and, indeed, the appearance of such branches suggests that they might have been partially burned by fire.

Causal Organism

Eutypa lata, teleomorph of an ascomycete (anamorph *Libertella blepharis* A.L. Smith), produces perithecial stromata on the surface of diseased sapwood two or more years after branch death. The stromata are obvious after the bark has fallen away; they are black and continuous, and locules of the perithecia may be seen when a thin slice is cut from the surface with a sharp blade (Plate 46, Fig. 8). The asci measure 40–60 × 5–7 μm, with pedicels 70–120 μm long. Ascospores (6.2–11 × 1.5–2 μm) are pale yellow, allantoid, thin-walled, and aseptate (Fig. 8).

E. lata may be cultured by aseptically transferring small chips of infected sapwood to potato-dextrose agar, malt agar, or other common laboratory media. White mycelium grows from infected chips after three or four days at 20–25°C. The fungus is a virulent pathogen but is not a strong competitor against other fungi. Its presence in a diseased branch sometimes may be difficult to detect if other fungi have also invaded the wood.

Perithecia are not produced in culture, but after six to eight weeks pycnidia sometimes develop, often exuding single-celled conidia (20–35 × 1–1.3 μm) (Fig. 9) in cream-colored or orange cirri or in translucent masses. Not all isolates sporulate in culture, and they vary in the amount of dark pigment produced in the medium after one or two weeks. Identification may be difficult for those unaccustomed to the appearance of the fungus in culture, and colonies five or six days old should be compared with a reliable reference culture transferred at the same time. The conidia do not germinate and apparently have no role in dissemination of the pathogen. The so-called "conidia" may in fact be spermatia.

Disease Cycle and Epidemiology

In semiarid regions such as California, southeastern Australia, Greece, or the Cape region of South Africa, development of the perithecial stromata is restricted to zones that receive mean annual rainfall of at least 350 mm. Where winters are relatively mild, as in the places just mentioned, there is a well-defined cycle of ascospore production from perithecia that mature late in the winter, with some regional variation. In California, spore production is abundant in mid-November, decreases from December through mid-January, and increases again from mid- or late January when new perithecia mature. In southern Australia, many ascospores are released with every rain exceeding about 2 mm from the time the new perithecia mature in late winter (mid- to late August) until the asci are finally exhausted in the following autumn (May). In contrast, where the winters are severe, as in the eastern United States, the potential for ascospore production is greatest in late winter and spring and least in summer.

Ascospores are disseminated by wind, and when deposited on parts of the host that spores cannot invade directly, rain showers may disperse them to the open ends of vessels exposed at fresh wounds, where they lodge at a depth of 2–4 mm below the surface. Once within a vessel, the ascospores germinate in 11 or 12 hr at the optimum temperature of 20–25°C. Mycelia proliferate slowly within vessels and later grow into adjacent elements of the functional wood, leading eventually to the appearance of a canker and, later, to death of the branch.

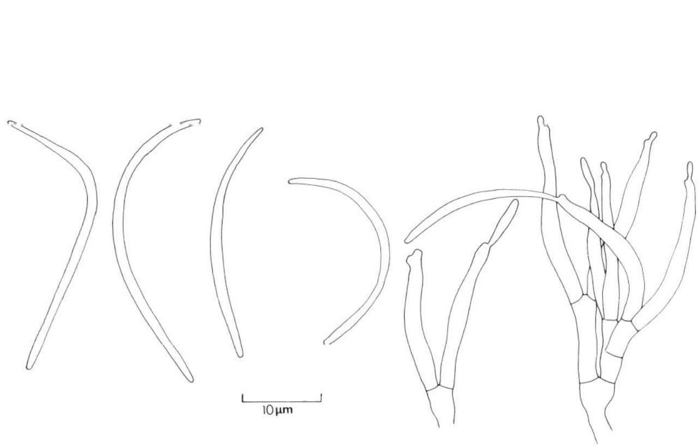

Fig. 9. Single-celled conidia of *Libertella blepharis,* the anamorph of *Eutypa lata.* (Reprinted, by permission, from Carter, 1991)

Management

Sanitation continues to be a sound practice, but growers cannot expect it to be effective if practiced only on their own property. Inoculum on other hosts and in neighboring orchards, vineyards, and gardens can nullify sanitation efforts.

All cultivars of apricot are susceptible to invasion by the pathogen. Susceptibility of plums seems to be restricted to the European cultivars (*P. domestica* L.).

E. lata depends entirely on wounds to gain access to vascular tissues. Pruning wounds are very important, and reduced amounts of pruning are beneficial. In certain regions where the summers are dry (e.g., the Central Valley of California), trees should be pruned soon after harvest. Trees subjected to mechanical topping and hedging are at great risk unless pruned when the risk of ascospore dispersal is minimal. In Switzerland, the incidence of disease in summer-pruned trees was reduced 75–90% compared with those with traditional winter pruning. Growers must become regular and astute observers of their trees and remove each infection as soon as it is detected. Pruning wounds may be protected with a fungicidal dressing when inoculum is present. Benzimidazole chemicals have been effective when applied as a fungicidal wound dressing, and pneumatic-powered pruning equipment has been modified to give the operator discretion to flood each pruning wound with the dressing. Conventional spray application of benzimidazoles, however, is ineffective because an insufficient amount of the chemical reaches the infection court within the vessel.

A psychological impasse often limits acceptance by growers of the need to begin routine protection when their trees are young and healthy. This routine must be faithfully maintained long before the effects of the pathogen become apparent.

Selected References

Bolay, A., Spring, J. L., and Monney, P. 1990. Influence de l'époque de la taille des abricotiers sur le développement de chancres. Rev. Suisse Vitic., Arboric., Hortic. 22:371-376.

Carter, M. V. 1991. The Status of *Eutypa lata* as a Pathogen. Phytopathological Paper No. 32. International Mycological Institute, London.

Carter, M. V., and Perrin, E. 1985. A pneumatic-powered spraying secateur for use in commercial orchards and vineyards. Aust. J. Exp. Agric. Anim. Husb. 25:939-942.

English, H., and Davis, J. R. 1978. *Eutypa armeniacae* in apricot: Pathogenesis and induction of xylem soft rot. Hilgardia 46:193-204.

Ogawa, J. M., and English, H. 1991. Diseases of Temperate Zone Tree Fruit and Nut Crops. Publ. 3345. University of California Division of Agriculture and Natural Resources, Oakland. 461 pp.

(Prepared by M. V. Carter)

Fungal Gummosis

Peach tree fungal gummosis was observed in the southeastern United States initially during the 1960s. About the same time, the disease was independently discovered in Japan, where it is described as peach blister canker or *ibokawa byo.* More recently, it was reported in China and Australia. This disease, characterized by symptoms associated with lenticels of the bark, contributes to a general decline of trees.

Symptoms

The earliest symptoms appear on young bark of vigorous trees as blisters 1–6 mm in diameter, generally each with a lenticel at its center (Plate 47). Blisters may be observed late in the season when infection occurs or the following spring. Beneath the outer bark, diseased tissue at the lenticel margin is surrounded by hyperplastic tissue (Plate 48). By the end of the second season, the area of necrosis surrounding lenticels has enlarged, and the hyperplasia is often less visible or absent. Some necrotic lesions exude resins. Lesions that appear in the second season after infection may or may not be preceded by the formation of blisters at the same infection site. Beginning when trees are two or three years old, sunken necrotic lesions encircling lenticels can be seen on the trunk and major branches. Typically, copious resin exudate is associated with lesions at multiple sites. Lesions 2 cm or more in diameter on the oldest bark may coalesce to form large cankers. Phloem and cortex are primarily affected; however, necrosis may extend to the xylem.

Peach orchards that are poorly managed or water-stressed are more likely to have severely infected trees. In its severest form, the disease kills branches or entire trees. When growing conditions are optimal and inoculum levels are low, a periderm is formed, and the diseased tissue peels away from the tree. After repeated infections, the bark becomes rough and scaly.

Causal Organisms

The symptoms just described probably are caused by a physiologic race of *Botryosphaeria dothidea* (Moug.:Fr.) Ces. & De Not. (syn. *B. berengeriana* De Not., anamorph *Fusicoccum aesculi* Corda) specific to peach. Other *Botryosphaeria* species have been reported to cause peach gummosis in the United States; however, these are primarily wound invaders that are not known to cause infections at lenticels. *B. rhodina* (Cooke) Arx is relatively rare on peach. *B. obtusa* (Schwein.) Shoemaker is very common in the southeastern United States, but its importance as a peach pathogen is unclear. *B. obtusa* is localized in dead tissue and outer bark (rhytidome) and is absent or infrequent in the newly infected cortex and phloem.

Stromata bearing pycnidia or perithecia of *B. dothidea* develop in dead bark and in lenticels of diseased bark. The perithecia, which are frequent in the United States but rare in Japan, contain clavate asci averaging 120 × 19 µm separated by pseudoparaphyses. Each ascus contains eight hyaline, nonseptate, fusiform ascospores measuring 17–28 × 9–12 µm. The pycnidia contain hyaline, nonseptate, fusiform conidia (15–29 × 5–8 µm). Microconidia are often produced in locules contained in pycnidial or ascogenous stromata.

B. dothidea grows well on Waksman medium, prune agar, or potato-dextrose agar. Carbon sources for growth are glucose, fructose, sucrose, or starch. As a nitrogen source, sodium or potassium nitrate supports good growth. The fungus grows at 20–36°C (optimum 27°C) in a medium with pH 4–7. At first, the culture forms white mycelial growth, which changes to grayish white and later grayish black.

Disease Cycle and Epidemiology

B. dothidea overwinters in diseased bark and in dead and dying wood, where it sporulates profusely. The fungus spreads within the orchard mainly by dispersal of conidia in rainwater. Where present in the Northern Hemisphere, conidia are available during March through October. Infections at lenticels occur from March through August, with most infection during May through July in the United States and mid-June to late July in Japan. The fungus also invades through wounds. Cankers that develop from artificially inoculated wounds remain active more than one year and lead to secondary infections at lenticels. Blossoms, leaves, and fruit are not infected.

Control

Dead wood should be removed during winter pruning, and prunings should be taken from the orchard or destroyed mechanically (e.g., with a flail mower) to reduce inoculum from these sources. Summer pruning should be avoided, because available inoculum can germinate and colonize wounds rapidly at this time. Also, alleviating water and nutrient stress can reduce disease incidence and severity. Fungicides such as captafol (no longer sold in the United States) and captan, applied repeatedly and thoroughly to bark before and during the infection period, are effective for prevention but may not be economically feasible in many peach-growing areas. Spread of the disease has also been reduced in China by first removing infected bark and then treating with these fungicides.

Selected References

Abiko, K., and Kitajima, H. 1970. Blister canker, a new disease of peach tree. Ann. Phytopathol. Soc. Jpn. 36:260-265.

Chen, X. Z. 1985. Studies on the gummosis of peach (*Prunus persica*) caused by *Botryosphaeria dothidea*. Acta Phytopathol. Sin. 15:53-57.

Koganezawa, H., and Sakuma, T. 1984. Causal fungi of apple fruit rot. Bull. Fruit Tree Res. Stn. Ser. C (Morioka) 11:49-62.

Pusey, P. L. 1989. Availability and dispersal of ascospores and conidia of *Botryosphaeria* in peach orchards. Phytopathology 79:635-639.

Weaver, D. J. 1979. Role of conidia of *Botryosphaeria dothidea* in the natural spread of peach tree gummosis. Phytopathology 69:330-334.

Wu, Y. 1985. Studies on pathogen of peach blister canker [*Physalospora persicae* (Abiko et Kitajima)]. Inst. Hortic. Shanghai Acad. Agric. Sci.

(Prepared by P. L. Pusey, H. Kitajima, and Y. Wu)

Black Knot

Black knot can be very destructive on plums and prunes in home plantings and is common on wild plum and cherry trees. Black knot was first described in 1821 in Pennsylvania and was considered one of the most destructive diseases of plum and tart cherry in the northern states during the late 1880s. Losses today are usually not extensive in commercial orchards except where the disease has become established. Once established on susceptible cultivars, black knot is very difficult to control, requiring pruning of affected branches and application of fungicides. This fungal disease occurs occasionally on apricot, peach, sweet and sour cherry, and ornamental as well as wild *Prunus* species. Plum cultivars differ in susceptibility, but greater differences in severity are found among species of cultivated and wild plum. Black knot is distributed throughout North America but is most severe in the eastern and northern fruit-growing regions.

Symptoms

The disease affects only woody parts of trees, primarily twigs, and is characterized by elongated swellings 1.5–30 cm long and 2.5–7.5 cm in circumference. The knots seldom surround the entire limb, but numerous infections on twigs or small branches may cause general decline in growth and lead to loss of branches and tree death in a few years. When first formed, knots are olive green and firm but rather corky. Later, they turn black and become hard and brittle (Plate 49). Twigs affected near the tip may become bent at right angles, but in general the knots are difficult to detect until the following year. Old knots expand lengthwise annually, producing new elongated swellings at each end.

Causal Organism

Apiosporina morbosa (Schwein.:Fr.) Arx (syn. *Dibotryon morbosum* (Schwein.:Fr.) Theiss. & Syd., anamorph *Fusicladium* sp.), the cause of black knot, is an ascomycetous fungus belonging to the order Dothideales and family Venturiaceae. The fungus produces ascostromata composed of thick-walled, pigmented, rounded-angular cells on the surface of hypertrophied host tissue. Ascospores produced in clavate asci, 120 µm long, and located in locules in the ascostromata, are two-celled (one smaller than the other), ovate, light green or hyaline, measuring 12–16 × 8–10 µm. Conidia (3–20 × 2–5.5 µm) are ovate, single-celled, and light brown and are produced at the apex of broad, flexuous, septate, brown conidiophores, which arise from the outer layer of the stromatic cells.

Disease Cycle and Epidemiology

Infections occur on new shoot growth, mainly from ascospores released when perithecia are wet, during periods of measurable rainfall of 6 hr or more at 21°C. Ascospore germination is very low at 6°C but increases significantly at 12, 18, and 24°C. Very little infection is known to occur from conidia. Susceptible twigs may be infected soon after budbreak and throughout active shoot elongation, but most infections seem to occur just before and during bloom or after petal fall, depending on location. The fungus may penetrate nonwounded surfaces. Hyphae grow inward to the cambial layer and later along the medullary rays in the xylem of the twig. Excessive parenchyma cells are produced and are pushed outward to form the knot. The first symptoms are visible by early autumn, but further development continues the following spring. The knots develop rapidly the second summer, as do the ascostromata, in which ascospores are formed. Ascospores mature during the second winter after infection is initiated.

Control

Plum cultivars resistant to black knot should be selected where possible. Cultivar susceptibility among *Prunus* species ranges from highly susceptible to nearly immune. Among the cultivars grown east of the Rocky Mountains, Stanley, Damson, Bluefree, and Shropshire are considered the most susceptible; Fellenburg, Methley, Milton, Bradshaw, and Early Italian are moderately susceptible; Formosa, Shiro, and Santa Rosa are slightly susceptible; and President is highly resistant.

Black knot control is more effective when inoculum sources such as wild plums and cherries are removed from nearby orchard sites. Knots on small twigs and branches should be pruned 8–10 cm below the visible swelling and removed from orchards. Infected tissue on large limbs or trunks should be removed at least 2 cm beyond the visible swelling; removal is more effective if done in midsummer, when visible swelling is close to the advancing fungal growth in the stem. Black knot can be controlled by fungicides such as captan, chlorothalonil, and benzimidazoles applied during active shoot elongation.

Selected References

Gourley, C. O. 1962. A comparison of growth, life cycle, and control of *Dibotryon morbosum* (Schw.) Theiss. and Syn. on peach and plum in Nova Scotia. Can. J. Plant Sci. 42:122-129.

Koch, L. W. 1935. Investigations on the black knot of plums and cherries, IV. Studies in pathogenicity and pathological histology. Sci. Agric. 15:729-743.

Ritchie, D. F., Klos, E. J., and Yoder, K. S. 1975. Epidemiology of black knot of Stanley plums and its control with systemic fungicides. Plant Dis. Rep. 59:499-503.

Rosenberger, D. A., and Gerling, W. D. 1984. Effect of black knot incidence on yield of Stanley prune trees and economic benefits of fungicide protection. Plant Dis. 68:1060-1064.

Smith, D. H., Lewis, F. H., and Wainwright, S. H. 1970. Epidemiology of the black knot disease of plums. Phytopathology 60:1441-1444.

Wainwright, S. H., and Lewis, F. H. 1970. Developmental morphology of the black knot pathogen on plum. Phytopathology 60:1238-1244.

(Prepared by K. D. Hickey)

Ceratocystis Canker

This disease, first described on almond (*Prunus dulcis*) as "mallet wound canker," has existed in California almond and prune orchards for many years. Cankers develop in bruises made on the bark of the trunk and scaffold branches by harvesting or cultivation equipment. In the late 1950s, the disease assumed major importance on prune and almond, when mechanical harvesting operations were developed. In recent years, improvements in harvesting equipment and closer attention to other epidemiological factors have reduced the importance of this disease. Ceratocystis canker is known only in California and may be found on apricot and peach.

Symptoms

Diseased bark becomes water-soaked, darkened, and sunken; amber-colored gum commonly exudes from the canker margin. The necrotic cortex is reddish brown in prune (Plate 50) and dark brown in other *Prunus* spp. A dark brown stain in the xylem may extend 50 cm or more beyond the bark canker margin. A limb or trunk 10–15 cm in diameter can be girdled and killed in three to four years. Removal of the loosened bark of a recently infected wound often reveals the whitish mycelium of the pathogen on the exposed cambium.

Causal Organism

Ceratocystis canker is caused by *Ceratocystis fimbriata* Ellis & Halst. (anamorph *Chalara* sp.), a diverse species comprising numerous strains characterized by pathogenic specialization and variable asexual structures. Teleomorph morphology, however, is relatively uniform. The globose, long-necked perithecia of California isolates measure 133–240 µm in diameter; the necks are 300–760 µm long, and their apical diameter is approximately 11 µm. The hat-shaped ascospores (2 × 6 µm) have a tapered, gelatinous brim and form within evanescent asci. Dimensions of the three types of asexual spore forms are as follows: cylindrical, unicellular, hyaline to pale brown endoconidia, 2.4 × 17.4 µm; doliform, hyaline to pale brown endoconidia, 6.4 × 7.2 µm; and doliform to pyriform, unicellular, pale to dark brown, thick-walled conidia (chlamydospores), 10.5 × 12.9 µm. Mean conidiophore dimensions are 4.8 × 79.8 µm, 6.1 × 34.9 µm, and 4.9 × 33.8 µm, respectively. At 25°C, cylindrical endoconidia germinate within 24 hr and the thick-walled conidia in 48 hr; optimum growth occurs at 24–27°C. Thiamine is required for perithecial production. The fungus is homothallic, but the ascospore progeny contain some self-sterile, cross-fertile strains. Carrot root disks selectively isolate the pathogen.

Disease Cycle and Epidemiology

C. fimbriata overwinters as mycelium in perennial cankers, as fruiting bodies and spores in old infected wounds, and in association with its insect (or mite) vectors. Arthropod vectors are so dominant in the disease cycle that if they are excluded from fresh wounds, no infection occurs. Major vectors are a nitidulid beetle (*Carpophilus freemani*) and a drosophilid fly (*Chymomyza procnemoides*). Several other insects and one mycophagous mite are secondary vectors. Insects are attracted to fresh bark wounds, where they deposit spores of the pathogen and then feed on the resultant fungal colonies. Viable spores are present in insect frass, and some adults that emerge from pupae of contaminated larvae harbor the pathogen.

Fresh bark injuries are highly susceptible to infection but become resistant within 14 days. Resistance is related to the deposition of lignin and suberin in the wound periderm. The fungus initially colonizes the exposed cambium and injured cortex and then invades uninjured tissues, including the outer xylem. The deeper, dark-stained xylem does not contain the pathogen. In older cankers, the fungus usually can be isolated only from canker margins. The cankers are perennial, developing rapidly during summer and slowly during cooler weather. Bark injured soon after an irrigation attracts more insect vectors and develops more infection than bark injured before irrigation.

Control

Since most injuries that lead to infection result from mechanical harvesting, shakers that cause minimal damage should be used. Irrigation should be avoided during the three weeks preceding harvest.

Severely cankered or dead limbs should be removed by cutting at least 15 cm below the canker margin. With small to medium cankers, removal of the infected bark and outer xylem is effective. Surgical treatments should be done during cool weather when vectors are inactive.

Selected References

Bostock, R. M., and Middleton, G. E. 1987. Relationship of wound periderm formation to resistance to *Ceratocystis fimbriata* in almond bark. Phytopathology 77:1174-1180.

DeVay, J. E., Lukezic, F. L., English, H., Trujillo, E. E., and Moller, W. J. 1968. Ceratocystis canker of deciduous fruit trees. Phytopathology 58:949-954.

Moller, W. J., and DeVay, J. E. 1968. Insect transmission of *Ceratocystis fimbriata* in deciduous fruit orchards. Phytopathology 58:1499-1508.

Moller, W. J., DeVay, J. E., and Backman, P. A. 1969. Effect of some ecological factors on Ceratocystis canker in stone fruits. Phytopathology 59:938-942.

Ogawa, J. M., and English, H. 1991. Diseases of Temperate Zone Tree Fruit and Nut Crops. Publ. 3345. University of California Division of Agriculture and Natural Resources, Oakland. 461 pp.

(Prepared by H. English)

Sclerotium Stem Rot

Young stone fruit trees in nurseries or during their first year of growth after transplanting may die from stem rot caused by *Sclerotium rolfsii* Sacc. The fungus invades stems near the soil line, develops a prominent white mycelium on the tissue surface, and later invades roots. Infected plants wilt and die quickly. Spherical sclerotia, white at first and later turning brown, form on the mycelial mat. The pathogen is favored by high temperatures, wet soil, and high levels of organic matter (e.g., sawdust). Shaded conditions that encourage high relative humidity should be avoided. Fungicide drenches have been used successfully to arrest the spread of the pathogen.

Selected Reference

Magee, C. J. 1947. Sclerotium stem rot caused by *Sclerotium rolfsii*. Agric. Gaz. N.S.W. 58:265-267.

(Prepared by E. I. Zehr)

Dieback

Dieback of sweet cherry trees, caused by *Phialophora parasitica* Ajello, L.K. Georg, & Wang, was a serious disease in Greece in the 1980s, resulting in death of many trees. The disease has not been reported on cherries elsewhere to our knowledge.

Symptoms

Twigs may fail to form leaves in spring, and later leaves on some branches wilt suddenly and abscise. Except for necrosis, no other visible symptoms appear on affected twigs, but the disease continues to spread downward on the branch to the trunk. The invasion may proceed upward and downward in the heartwood until the whole tree is affected and dies.

Causal Organism

P. parasitica grows slowly on potato-dextrose agar but forms two colony types—glabrous, velvety colonies that are white with light red to brown-red centers, or floccose, white colonies with yellow-gray to olive-gray centers as the colonies age. Conidia are hyaline, single, ovoid to allantoid, aggregated into rounded, slimy heads at apices of phialides, measuring 2.3–7.6 × 1.1–2.8 μm. The organism is distributed in North America, Central America, northern Africa, and the Middle East.

Disease Cycle, Epidemiology, and Control

Little is known of the disease cycle or of control practices that might be effective. An association with wood-boring beetles in affected trees was reported. *Quercus* sp. has been reported to be a host.

Selected Reference

Rumbos, I. C. 1986. *Phialophora parasitica*, causal agent of cherry dieback. J. Phytopathol. 117:283-287.

(Prepared by E. I. Zehr)

Rhodosticta Canker

Rhodosticta quercina J.C. Carter, the causal agent of Rhodosticta canker, caused rapid dieback and decline in an 11- to 12-year-old California plum (*Prunus domestica*) orchard from 1958 to 1963. Since the initial outbreak, no further occurrence of Rhodosticta canker has been reported.

Symptoms and Epidemiology

Symptoms are most apparent during the growing season, when branches and limbs of plum trees are girdled by cankers. Leaves turn brown and flag the presence of cankers. Perennial cankers that are severe on older trees often are associated with injuries resulting from sunburn or shot-hole borers (*Scolytus rugulosus* Ratz.) and are most active during late spring and summer. Typical symptoms are collapsed cortical tissue coincident with the canker, exudation of dark or orange-colored gum, and a zonate margin around the canker in the inner bark tissues. Orange pycnidia develop just beneath the bark periderm and give cankers an orange coloration. Following periods of rain and during humid conditions, amber-colored cirri (spore tendrils) exude from the pycnidia (Plate 51), and the conidia are rain-splashed or are carried by insects to branches stressed by sunburn or other factors. Cankers seldom if ever originate at pruning wounds.

Causal Organism

R. quercina is a deuteromycete in the family Xythiaceae. Its stromata are lighter in color than those of *Leucostoma persoonii* Höhn, which is the main cause of dieback of prune and plum trees in California. On agar and in bark tissues, *R. quercina* produces orange pycnidia and an orange pigment that diffuses through the medium and bark. Both species are highly pathogenic on President plum and, except for pigmentation, have similar pathogenic and pycnidial characteristics in bark tissues.

Control

Avoidance of sunburn on branches and limbs, either through careful pruning or by use of whitewash, has been the most effective approach for controlling this disease. Cultivars of *Prunus salicina* Lindl. (Japanese plum) are resistant.

Selected Reference

Lukezic, F. L., DeVay, J. E., and English, H. 1965. Comparative physiology and pathogenicity of *Leucostoma persoonii* and *Rhodosticta quercina*. Phytopathology 55:511-518.

(Prepared by J. E. DeVay)

Phomopsis Blight

Phomopsis blight of apricot, characterized by necrotic spots on stems and fruit, has been reported in Algeria, but the disease is not known to cause significant economic loss. Spots are dark brown to black, irregular, and dispersed and produce pycnidial stromata externally. Pycnidia produce alpha (ovoid) and beta (filiform, curved) conidia. The pathogen is *Phomopsis vexans* (Sacc. & Syd.) Harter (teleomorph *Diaporthe vexans* (Sacc. & Syd.) Gratz).

Selected Reference

Narendra, D. V., Nezzal, T., and Santiago, J. C. 1979. Phomopsis blight of apricot. FAO Plant Prot. Bull. 27:132-133.

(Prepared by E. I. Zehr)

Armillaria and Clitocybe Root and Crown Rots

All species of stone fruits and many other woody plants may be affected by Armillaria root rot. The disease is widespread among stone fruit production areas of the world, but losses have been greatest in production areas of North America. Armillaria root rot is common in orchards planted on land recently cleared of forest trees or when infested orchard sites are replanted. Death of many trees often leads to abandonment of the orchard site, because practices to control the disease are not effective.

Clitocybe root rot causes severe losses predominantly in the southeastern United States, ranging from Virginia and West Virginia to Florida and west to Oklahoma and Texas. Reports of its occurrence as far north as Michigan and New York need to be confirmed.

Symptoms

Trees infected when older than five years often exhibit poor terminal growth and undersized leaves. Foliage often remains green until midsummer, when the whole tree collapses with leaves attached (Plate 52). Sudden collapse of trees in midsummer, rather than a gradual yellowing of leaves followed by defoliation, is indicative of Armillaria root rot rather than Phytophthora root rot. Also, Armillaria root rot is more common on light, well-drained, sandy soils, whereas Phytophthora root rot is common on heavy, poorly drained, clay soils. Trees often die in a circular pattern from a focus of infection within an orchard.

A fan-shaped, white fungal mat often is present between the bark and the wood on the crown of affected trees (Plate 53). Dark brown to black threadlike rhizomorphs may be observed on the surface of the bark (Plate 53). Rhizomorphs are about the size of a shoestring and also are found on dead roots, on the surface of live roots in advance of the necrotic tissue, and in the soil around the roots of infected trees. Clusters of mushrooms may arise at the base of dead trees in autumn (Plate 54).

Symptoms of Clitocybe root rot are similar to those of Armillaria root rot, but the white fungal mat associated with Clitocybe root rot often has perforations in it. Rhizomorphs are not formed, but black, hardened mycelial extrusions develop in cracks in the bark. Yellow to yellow-brown mushrooms may develop at the base of stumps in the autumn (Plate 55). Unlike Armillaria root rot, Clitocybe root rot is often associated with clay or sandy clay soils.

Causal Organisms

Armillaria mellea (Vahl:Fr.) P. Kumm. in the broad sense causes Armillaria root rot. Mycelium of *A. mellea* differs from that of most hymenomycetes in that clamp connections are present only during a brief stage in the life cycle. Rhizomorphs are long, black, threadlike aggregations of mycelium. Gill-bearing mushrooms with white spores produced in clumps around dead stumps have a somewhat thickened base and a ring (annulus) attached to the stipe. A single fruiting body can produce basidiospores in four sexual compatibility groups (tetrapolar mating). Colonies produced in culture by single, haploid basidiospores produce fluffy, white mycelial growth, whereas colonies produced by diploid mycelium produce flat, often dark, crustose mycelial growth. *Armillaria* is the only hymenomycetous fungus currently known to produce a genetically stable diploid rather than dikaryotic mycelia.

Recent genetic studies with *Armillaria* indicate that *A. mellea* sensu lato is a complex of several reproductively isolated groups or biological species. Based on sexual incompatibility studies, four species of *Armillaria* were identified among isolates of the fungus recovered from trees of sour cherry, sweet cherry, and peach in Michigan; these are, in order of importance, *A. ostoyae* (Romagn.) Herink., *A. mellea* sensu stricto (Vahl:Fr.) P. Kumm., North American biological species III, and *A. bulbosa* (Barla) Kile & Watling.

A. tabescens (Scop.) Dennis, Orton & Hara (syn. *Clitocybe tabescens* (Scop.) Bres.) causes Clitocybe root rot. Its fruiting bodies lack an annulus but otherwise are morphologically quite similar to those of *A. mellea* sensu lato. Also, *A. tabescens* readily produces fruiting bodies under artificial conditions, whereas *A. mellea* fruits only rarely in culture. Its nuclear condition is identical to that of *A. mellea*, but incompatibility of isolates among the two species indicates that the species are biologically distinct.

Disease Cycle and Epidemiology

A. mellea infects many native, ornamental, and agricultural woody plants. *Armillaria* spp. may exist in an orchard site before planting, because mycelium may persist for many years on infected roots of removed forest trees or other tree species. In addition, rhizomorphs may survive in the soil or on dead, infested roots. Recently, it was estimated that a colony of *A. bulbosa* had existed in a Michigan site for at least 1,500 years.

Armillaria spreads within orchards by growth of rhizomorphs, fungal extrusions, and mycelial strands along roots, or possibly through the soil, to contact points with roots of adjoining trees. Observed rates of tree-to-tree spread of *A. mellea* in peach orchards in Australia ranged from 0.8 to 3.2 m per year. Because new infections are adjacent to previously infected trees or tree sites, the disease usually develops outward from centers of infection. Spread to new orchard sites is by movement of infested soil or infected roots. Basidiospores are not considered important in the disease cycle.

Control

Although no common species of stone fruit is immune to *Armillaria, Prunus* rootstocks vary in susceptibility. Species in the *Prunus* sections *amygdalus* (*P. persica, P. dulcis, P. mira,* and *P. davidiana*), *armeniaca* (*P. armeniaca*), and *cerasus* (*P. avium,* mazzard, and *P. mahaleb*) are very susceptible, whereas species in the sections *euprunus* (*P. domestica, P. insititia, P. cerasifera,* and *P. spinosa*) and possibly *prunocerasus* (*P. subcordata* and *P. munsoniana*) are moderately susceptible to *A. mellea.* In general, *Prunus* species are more susceptible than apple or pear. Relative resistance rankings may be complicated by interactions between different species of *Armillaria* and rootstock genotypes. Sweet cherry on mazzard rootstocks is more resistant to *A. mellea* than is sour cherry on mazzard rootstocks.

Newly cleared forest land or old orchard sites with a history of Armillaria or Clitocybe root rot should not be planted to stone fruit crops. If replanting of infested tree sites is attempted, the soil in the site should be worked well to remove as many old roots as possible. Fumigation has been only moderately effective in controlling the disease, because the fungus can survive deep in the soil and in roots that are not easily permeated by chemical vapors. Deep plowing or chiseling should not be attempted in affected areas.

Selected References

Anderson, J. B. 1983. Bifactorial heterothallism and vegetative diploidy in *Clitocybe tabescens*. Mycologia 74:911-916.

Anderson, J. B. 1986. Biological species of *Armillaria* in North America: Redesignation of groups IV and VII and enumeration of voucher strains for other groups. Mycologia 78:837-839.

Guillaumin, J. J., Pierson, J., and Grassely, C. 1991. The susceptibility to *Armillaria mellea* of different *Prunus* species used as stone fruit rootstocks. Sci. Hortic. (Amst.) 46:43-54.

Kable, P. F. 1974. Spread of *Armillariella* sp. in a peach orchard. Trans. Br. Mycol. Soc. 62:89-98.

Proffer, T. J., Jones, A. L., and Ehret, G. R. 1987. Biological species of *Armillaria* isolated from sour cherry orchards in Michigan. Phytopathology 77:941-943.

Rhoads, A. S. 1945. A comparative study of two closely related root rot fungi, *Clitocybe tabescens* and *Armillaria mellea*. Mycologia 37:741-766.

Rhoads, A. S. 1954. Clitocybe root rot found widespread and destructive in Georgia and South Carolina peach orchards. Plant Dis. Rep. 38:42-46.

Smith, M. L., Bruhn, J. N., and Anderson, J. B. 1992. The fungus *Armillaria bulbosa* is among the largest and oldest living organisms. Nature 356:428-431.

(Prepared by A. L. Jones)

Phytophthora Root and Crown Rots

Phytophthora root and crown rots are among the most important soilborne diseases that affect stone fruits. The diseases occur worldwide. Losses have been reported on all major stone fruits, and all ages of the trees can be affected. Phytophthora root and crown rots often result in the death of affected trees.

Phytophthora spp. were first associated with extensive losses of nursery and orchard stone fruit trees in the early 1900s in California. Over the last 30 years, the reported incidence of Phytophthora root and crown rots of stone fruit trees has increased—a result of improved techniques of isolation and study of *Phytophthora* spp. and of changes in horticultural practices during the period. For example, rootstocks for stone fruits are often selected without sufficient attention to resistance to soilborne pathogens. In addition, as fruit production has expanded, many orchards have been established on poorly drained soils that favor infection by *Phytophthora* spp. Spread of *Phytophthora* spp. within and between orchards and nurseries probably has been facilitated by machinery and irrigation water that are infested with the pathogens.

Symptoms

Aboveground symptoms of Phytophthora root or crown rot in stone fruit trees often appear first as insufficient extensions of shoot growth. If the root system is infected in the fall, aboveground symptoms may not be apparent until the following season. Leaves on affected trees usually become sparse, small, and chlorotic. Fruits may be undersized, highly colored, and sunburned. As the disease progresses, shoot and scaffold dieback may occur, often accompanied by sunburning of the exposed bark and secondary invasion by bark-boring insects. Trees often die within weeks or months of the first symptoms (Plate 56), but in other cases tree decline is gradual, extending over several growing seasons. Decline tends to be more rapid in nursery and young orchard trees than in large, well-established trees. Aboveground symptoms of decline in trees can be confused with other diseases and disorders that result from root or vascular dysfunction.

Trees with aboveground symptoms of decline caused by *Phytophthora* spp. may show symptoms of crown rot, which refers to decay on the root crown and/or the trunk base near the soil line. Bark of affected crown tissue is killed and turns various shades of brown. The discoloration may extend into outer layers of subtending xylem tissue. As the decay expands, a canker develops, often accompanied by balls or strands of gum that exude from affected bark. The cankers have distinct borders that become apparent when outer bark is scraped to reveal the edges of necrotic (various shades of brown) and healthy (white to greenish white) tissue. Exposed cankers sometimes show a zonate appearance resulting from various shades of discoloration in the necrotic tissue colonized in succession. The zonate characteristic apparently is uncommon on trees affected in midwestern and northeastern parts of the United States. Lateral expansion of cankers can girdle trees, resulting in eventual death. Cankers are often limited to subterranean crown tissue, but sometimes they extend vertically up the trunk and occasionally into scaffold branches.

Trees affected with Phytophthora root rot often have few feeder roots, and those roots remaining are often decayed, with dark brown to black discoloration in the cortex and stele. The affected roots are initially firm or brittle. Bark tissue of older structural roots shows decay similar to that described for diseased crown tissue (Plate 57).

Phytophthora spp. can cause aerial cankers when floods or sprinkler irrigations carry inoculum of the fungi to upper parts of trees and favor infection. Under wet conditions, nursery trees can develop "bundle rot" cankers in stem tissue after the trees are dug from the nursery. The cankers may become apparent after trees are dug for storage or when the trees are removed from bundles and planted.

Causal Organisms

Isolations of *Phytophthora* spp. from tissue recently infected by the pathogen are more successful than isolations from dried and cracked areas in old cankers or from dried roots that are usually colonized by saprophytic soil fungi. Isolation of *Phytophthora* spp. is facilitated by semiselective media such as $P_{10}VP$ (corn meal agar amended with pimaricin, vancomycin, and PCNB) and PARP (corn meal agar amended with pimaricin, ampicillin, rifampicin, and PCNB). Some workers have incorporated hymexazol to suppress *Pythium* and *Mortierella* spp., which are otherwise able to grow.

Phytophthora spp. can be isolated from infested soil by various techniques, such as baiting with firm, green, unblemished pear fruits placed partially submerged in flooded soil samples. Some *Phytophthora* spp. can be isolated directly from soil with dilution plate techniques.

Many widely distributed *Phytophthora* spp. have been implicated as root and crown pathogens of stone fruit trees (Table 1). They possess hyaline, coenocytic hyphae (variable, but typically 5–9 μm in diameter) that may develop cross walls as cultures age. In liquid culture, and sometimes on solid media, sporangia form and can germinate directly by means of a germ tube or indirectly by production and release of zoospores. Oospores form following the union of male antheridia and female oogonia. They are produced by homothallic species in single culture, whereas heterothallic species require two mating types (A1 and A2) and/or sometimes other stimuli before sex organs form. Several *Phytophthora* spp. form chlamydospores and hyphal swellings.

Diagnosticians should take care not to mistake closely related and commonly isolated *Pythium* spp. for *Phytophthora* spp. In contrast to *Pythium* spp., which differentiate zoospores in a vesicle outside of the sporangium, *Phytophthora* spp. differentiate zoospores within the sporangium. Maturation and release of zoospores can be observed in soil extract water or in defined mineral solutions. Hyphae of *Phytophthora* spp. typically have wider branching angles and tend to be more coarse and rigid in appearance than those of *Pythium* spp.

Some key morphological characteristics of individual *Phytophthora* spp. that affect stone fruits are listed below, but these descriptions of necessity are abbreviated. Reliable identification of *Phytophthora* spp. can be difficult, often requiring the use of specialized media and techniques.

P. cactorum (Lebert & Cohn) J. Schröt. produces sporangia and sex organs abundantly in single cultures on solid media. Sporangia are broadly ellipsoid or ovoid to obpyriform (36–50 × 28–35 µm), conspicuously papillate, sympodial, nonproliferous, and deciduous, with a short pedicel (up to 4 µm long). Oogonia are spherical (19–38 µm), tapering at the base; oospores are aplerotic. Antheridia are paragynous. Some isolates form chlamydospores. On corn meal agar, cultures are slightly radiate to uniform, with very little aerial mycelium.

P. parasitica Dastur in liquid and on solid media produces sporangia that are broadly ovoid or pyriform to spherical (approximately 38 × 30 µm), papillate, sometimes intercalary, and deciduous, with a short pedicel (2 µm long). Chlamydospores form on solid media. As cultures age, many isolates produce sex organs in single culture; oogonia are usually less than 30 µm in diameter, with amphigynous antheridia. On corn meal agar, cultures can vary, but they commonly exhibit a characteristic "bushy" and irregular rosette pattern.

P. citrophthora (R.E. Sm. & E.H. Sm.) Leonian produces papillate sporangia that form most rapidly and abundantly in soil extract solution, although with time they can form on solid media. Sporangia are highly variable in size and shape and often possess more than one apex. Sex organs have not been observed in single cultures. On corn meal agar, cultures are finely radiate, often with flame-shaped sectors.

P. citricola Sawada forms sporangia much more rapidly and abundantly in liquid culture than on typical solid media. Sporangia are semipapillate; obpyriform, obovoid, sometimes ellipsoid; highly variable in shape and directional relationship of central axis to sporangiophore; often with two widely divergent apices; often sympodial, nonproliferous; and nondeciduous. Sex organs are produced abundantly on most media; oogonia are 17–44 µm in diameter. Antheridia are paragynous and attached to the oogonium at any point on the circumference. On corn meal agar, cultures are finely radiate, often with narrow lanceolate sectors.

P. syringae (Kleb.) Kleb. is a low-temperature species (optimum for growth near 20°C, maximum near 24°C). Sporangia will form in soil extract solution; they are broadly ovoid or obpyriform (average 57 × 36 µm), scarcely papillate to semipapillate, and nondeciduous. Sex organs are produced in single culture; oogonia are spherical (average 33 µm in diameter), and antheridia are predominantly paragynous. On corn meal agar, cultures usually grow slowly in a rosette pattern.

P. megasperma Drechs. does not produce sporangia on solid media but forms them readily in soil extract solution. They are ovoid (typically 35–50 × 25–35 µm), nonpapillate, and nondeciduous. Sex organs are produced on solid media, but formation is often delayed on rich media (up to several weeks). Antheridia are predominantly paragynous. Both "large" and "small" oospore types have been isolated from stone fruits (oogonia 16–61 µm).

P. cambivora (Petri) Buisman, *P. cinnamomi* Rands, *P. cryptogea* Pethybr. & Lafferty, and *P. drechsleri* Tucker are all heterothallic and produce nonpapillate, nondeciduous sporangia in soil extract solution. On solid media, *P. cryptogea* and *P. drechsleri* will form a few sporangia with delay, but *P. cambivora* and *P. cinnamomi* form none. Sporangia are ellipsoid to ovoid for *P. cinnamomi* (typically 57 × 33 µm), ovoid for *P. cambivora* (55–65 × 40–45 µm), ovoid to obpyriform for *P. cryptogea* (average 37–40 × 23 µm), and ovoid, obpyriform, or ellipsoid for *P. drechsleri* (sporangial dimensions are typically similar to those of *P. cryptogea*, but base of sporangium has a longer taper). When isolates of *P. cambivora*, *P. cinnamomi*, *P. cryptogea*, or *P. drechsleri* are paired with isolates of the opposite mating type, sex organs will often form. In such pairings, oogonial walls of *P. cambivora* exhibit characteristic bullate protuberances, whereas similarly produced oogonial walls of *P. cinnamomi*, *P. cryptogea*, and *P. drechsleri* are smooth. Antheridia are amphigynous with all four species; they are often two-celled in pairings with *P. cambivora* and *P. cinnamomi* but are one-celled with *P. cryptogea* and *P. drechsleri*. Hyphal growth of *P. cambivora* will exhibit microscopically a geniculate pattern, especially in lateral branches; hyphae of *P. cinnamomi* are uniquely coralloid. On corn meal agar, cultures are uniform with *P. cambivora*, somewhat rosette with *P. cinnamomi*, and somewhat radiate (often with narrow sectors) with *P. cryptogea* and *P. drechsleri*.

Disease Cycle and Epidemiology

Exposure of stone fruit trees to inoculum of *Phytophthora* spp. is common. Several *Phytophthora* spp. may be endemic in many temperate soils, including *P. cactorum*, *P. megasperma*, and *P. syringae*. Inoculum also may be introduced into orchard sites via infested soil, water, or plant material. Irrigation water from rivers, canals, and ponds can be contaminated with *Phytophthora* spp.

When conditions allow, infection is initiated in root, crown, trunk, or scaffold tissues. Evidence suggests that zoospores are usually the principal agents of infection, but other propagules

TABLE 1. Species of *Phytophthora* Causing Root and Crown Rots of Selected Stone Fruit Trees and Their Rootstocks

Species of *Phytophthora*	Known Distribution	Apricot Cultivars and Rootstocks	Peach Cultivars	Japanese Plum Cultivars and Rootstocks	Myrobalan Plum Cultivars	Prune Cultivars and Rootstocks	Sweet Cherry Cultivars and Rootstocks	Sour Cherry Cultivars and Rootstocks
P. cactorum	Temperate	+	+	+	+	+	+	+
P. cambivora	Cosmopolitan	+	+	+	+	+	+	+
P. cinnamomi	Cosmopolitan	+	+	–	+	–	+	–
P. citricola	Cosmopolitan	+	+	–	–	–	+	–
P. citrophthora	Cosmopolitan	+	+	–	–	–	+	–
P. cryptogea	Temperate	+	+	–	–	+	+	+
P. drechsleri	Cosmopolitan	+	+	–	–	+	+	–
P. megasperma	Cosmopolitan	+	+	–	–	+	+	+
P. parasitica	Warm regions	–	+	–	–	–	–	–
P. syringae	Temperate	+	+	–	–	+	+	+
Phytophthora sp.	Temperate	+	+	–	–	+	+	–

may infect also. Aerial portions of orchard and nursery stone fruit trees are subject to infection when inoculum reaches them in infested sprinkler water, windblown or splashed droplets or soil particles, or floodwater. *Phytophthora* spp. can infect stems of bare root nursery trees after digging if held in infested healing beds under wet conditions.

Development of Phytophthora root and crown rots in stone fruits is influenced by the susceptibility of the host, the virulence of the species of *Phytophthora,* and ambient environmental conditions. Controlled studies indicate that peach trees are subject to seasonal variations in susceptibility to *P. citricola,* with greatest susceptibility to root, crown, and stem colonization during the growing season and lowest susceptibility during tree dormancy. However, the reverse may be true for *P. syringae* in stone fruit trees. Temperatures conducive to development of root and crown rots caused by individual *Phytophthora* spp. correspond roughly to those that favor production and indirect germination of the sporangia.

Incidence and severity of Phytophthora root and crown rots are closely linked to soil moisture conditions. Episodes of soil saturation stimulate production, release, and motility of zoospores by *Phytophthora* spp., but very slight soil water tensions can dramatically reduce or prevent release and motility of zoospores. The duration of periods of soil saturation can greatly influence severity of Phytophthora root and crown rots, but effects depend on rootstock and species of *Phytophthora* present. For example, in controlled greenhouse experiments, *P. megasperma* caused very little root or crown rot in mahaleb and mazzard cherry and Lovell peach when soil was saturated for 24 hr or less, but root and crown rot was severe when periods of soil saturation were extended to 48 hr. In contrast, *P. cambivora* in cherry and *P. cinnamomi* in peach caused significant root and crown rot even in the absence of soil saturation, and relatively short durations of soil saturation resulted in severe disease.

Between episodes of dispersal and infection, *Phytophthora* spp. can survive temporarily as mycelium in host tissue. Resilient oospores and chlamydospores can enable some *Phytophthora* spp. to persist in soil or host tissue for several years.

Control

Strategies for control of Phytophthora root and crown rots include careful soil water management, genetically resistant rootstocks, and fungicides. Biological control strategies have a future role, but to date, prudent water management and rootstock selection are the most effective and economical controls.

Minimizing the frequency and duration of periods of soil saturation, especially in the vicinity of root crowns, helps prevent lethal infection by *Phytophthora* spp. Ideally, orchards should be established on well-drained soils. Achievement of adequate water drainage may require surface leveling, installation of drainage conduits, or both. Trees may be planted on raised berms to facilitate drainage of water away from tree trunks and root crowns. Irrigation water emitters and sprinklers should be placed or directed to avoid deposition and accumulation of water in tree crotches where scaffold branches join, on trunks, and around root crowns. Irrigation should be scheduled to avoid periods of soil saturation that are longer than 24 hr. Trees should be irrigated according to need, because excessive irrigation increases the risk of infection by *Phytophthora* spp.

Rootstocks for stone fruits differ in resistance to *Phytophthora* spp. Based on field observations, laboratory isolations, and greenhouse studies, losses due to *Phytophthora* spp. are more likely in apricot trees grafted on stocks of apricot or Lovell peach than on stocks of Nemaguard, Myrobalan plum, or Marianna 2624; in cherry trees on stocks of mahaleb than on stocks of mazzard, Stockton Morello, or Colt; or in plum and prune trees on stocks of Lovell peach or domestic plum than on stocks of Myrobalan plum, Myrobalan 29 C, or Marianna 2624. When trees are planted, the graft union between rootstock and scion should stand several inches above the soil line, because some scions, such as those of commercial prunes, are much more susceptible to infection than are common rootstocks, such as Myrobalan plum.

Chemical control of Phytophthora root and crown rots has been difficult and expensive compared with other control strategies. Although suppression is temporary, methyl bromide-chloropicrin mixtures have been used successfully as fumigants to reduce populations of *Phytophthora* spp. before orchard establishment. After planting, the systemic fungicides metalaxyl (applied as a soil drench) and fosetyl-Al (applied as a foliar spray) have in some cases provided a level of protection.

Selected References

Haygood, R. A., Graves, C. H., and Ridings, W. H. 1986. Phytophthora root rot and stem canker of peach trees in Mississippi. Plant Dis. 70:866-868.

Mircetich, S. M. 1981. Rootstock diseases of prune trees. Pages 107-113 in: Prune Orchard Management. D. E. Ramos, ed. Agricultural Science Publications, University of California, Berkeley.

Mircetich, S. M. 1982. Phytophthora root and crown rot of apricot trees. Acta Hortic. 121:385-396.

Mircetich, S. M., and Keil, H. L. 1970. *Phytophthora cinnamomi* root rot and stem canker of peach trees. Phytopathology 60:1376-1382.

Mircetich, S. M., and Matheron, M. E. 1976. Phytophthora root rot and crown rot of cherry trees. Phytopathology 66:549-558.

Stamps, D. J., Waterhouse, G. M., Newhook, F. J., and Hall, G. S. 1990. Revised tabular key to the species of *Phytophthora*. Mycol. Pap. 162. Commonwealth Mycological Institute, Kew, Surrey, England. 28 pp.

Wilcox, W. F., and Ellis, M. A. 1989. Phytophthora root and crown rots of peach trees in the eastern Great Lakes region. Plant Dis. 73:794-798.

Wilcox, W. F., and Mircetich, S. M. 1985. Effects of flooding duration on the development of Phytophthora root and crown rots of cherry. Phytopathology 75:1451-1455.

(Prepared by G. T. Browne and S. M. Mircetich)

Verticillium Wilt

Verticillium wilt of stone fruit, also called blackheart, verticilliosis, Verticillium hadromycosis, and vascular wilt, occurs in many parts of the world and causes serious economic loss. The disease was first shown to be caused by a fungus in 1916 by Czarnecki, who later placed the pathogen in the genus *Verticillium.* Stone fruit crops commonly affected by this disease are apricot, peach, plum, sweet and sour cherry, and nectarine. In addition, stone fruit rootstocks commonly used in stone fruit production are susceptible to Verticillium wilt. These include Myrobalan plum, mahaleb cherry, *Prunus mume,* and *P. davidiana.*

Symptoms

On apricot and almond, initial symptoms appear as a sudden wilting of leaves on one or more branches in early summer. Leaves become dull in appearance and rapidly tan and wither. Often the leaves remain attached, resulting in a one-sided, flagged appearance. At times, symptoms develop on leaves at the base of branches and progress upward until the entire branch is affected. Young trees may be killed by this pathogen, but affected branches of older trees may leaf out the following year, only to be attacked again. The result of repeated wilting is generally stunted terminal growth and low productivity or death.

Verticillium wilt of cherries is common in Europe and North America. The initial symptoms on sweet cherry are a withering of leaves on one or more spurs on one-year-old wood (Plate 58). Trees become unthrifty, leaves are dulled and stunted, and fruit is abnormally small. Unlike other stone fruit, even older cherry trees are susceptible to wilt and generally do not recover from the disease.

On peach, Verticillium wilt can be a serious disease in young trees (Plate 59) and is known to occur in all major production areas of the United States, Canada, the Mediterranean Basin, Bulgaria, and New Zealand. However, some cultivars show almost complete recovery with age.

Verticillium wilt occurs in plums, but as a group, these are the least affected. Symptoms (darkening of the wood) are similar to those on other stone fruit (Plate 60).

On all species, sapwood of affected branches contains streaks of brown to black discoloration, but these are less distinct on sweet cherries.

Causal Organism

There has been some controversy as to the nature of the species of *Verticillium* involved in vascular wilt of stone fruit as well as other perennial hosts. *Verticillium dahliae* Kleb. is the accepted name for the pathogen causing wilt in both annual and perennial hosts. Previously the causal organism of Verticillium wilt of stone fruit was thought to be *V. albo-atrum* Reinke & Berthier.

In culture, *V. dahliae* initially produces prostrate, hyaline mycelium, which becomes flocculose and white, whitish to cream in reverse (underside) after one week, and which gradually darkens with the formation of black, thick-walled microsclerotia. Mature cultures, when viewed in reverse, are brownish cream to black after two to three weeks as a result of formation of dark resting mycelium. White sectors are frequent in otherwise gray colonies.

Conidiophores of *V. dahliae* are abundant, erect, hyaline, and verticillately branched, with three to four phialides (16–35 × 1–2.5 µm) arising at each node. Phialides are sometimes secondarily branched. Conidia (2.5–8 × 1.4–3.2 µm), which arise singly at phialide apices, are ellipsoidal to irregularly subcylindrical, hyaline, and mainly one-celled but occasionally one-septate. In contrast, *V. albo-atrum* produces conidiophores that are darkened at the base, with phialides 20–30 (–50) × 1.4–3.2 µm. Conidia, though similar in shape, are generally larger, measuring 3.5–10.5 (–12.5) × 2–4 µm. Dark resting mycelium is produced only in association with microsclerotia in *V. dahliae*. Microsclerotia are variable in shape and size, range from 15 to 50 (to 100) µm in diameter, and arise from a single hypha by repeated budding. In *V. albo-atrum*, dark brown to black, septate resting mycelium is formed after 10–15 days. Swelling between the septa of resting mycelium results in a torulose appearance, but microsclerotia are not produced. Distinction of the two species is based on the fact that *V. dahliae* grows at 30°C, whereas *V. albo-atrum* does not.

Disease Cycle and Epidemiology

V. dahliae is a soil inhabitant that can survive at depths up to 90 cm, but most microsclerotia are found in the upper 15–30 cm. Survival in soils is poor when temperatures are above 25°C, whereas flooding also reduces viability. However, survival of microsclerotia for several months to years is common when soil moisture is favorable, even at temperatures of up to 30°C.

The fungus can be disseminated in many ways, including by diseased plants that are transplanted, windblown soils, and propagation knives. Following introduction into soil, the pathogen may persist as dormant sclerotia or may increase sclerotial numbers by infecting other crops or weed hosts. In fallowed sandy and clay soils, inoculum may also decrease; however, the pathogen usually persists for extended periods.

Disease increase is more severe in wet soil. Disease development occurs at 21–27°C (optimum 24°C). Soil pH has little effect on disease development at values ranging from 4.4 to 6.7. In some crops, nutrient deficiency exacerbates disease.

Infection occurs early in the spring and is followed by symptom expression in midsummer. Although symptoms may become more severe in late summer, they generally do not appear for the first time during this period. The pathogen may survive from season to season in roots of infected trees.

In apricots, as infected wood becomes darkened, the pathogen may be inactivated. Extracts from discolored wood of branches have inhibited spore germination and hyphal growth. Gums and tyloses in the vessels prevent upward movement of conidia, and trees recover when new wood is formed.

Pathogenesis is poorly understood, but substances produced in culture that may be important include pectolytic and cellulolytic enzymes, a lipopolysaccharide, high- and low-molecular-weight proteins, and growth regulators, such as indole-3-acetic acid and ethylene.

Control

Verticillium wilt of stone fruits can best be controlled by planting in soil that has low or undetectable inoculum densities and that has been cropped to various grasses or other nonhosts, as there are no resistant rootstocks. Soil fumigation with chloropicrin or with methyl bromide and chloropicrin destroys inoculum of *V. dahliae*. Proper fertilization and irrigation of orchards may enable trees to recover from infection.

Selected References

Hanison, A. F., and Clare, B. C. 1970. Host reactions involved in recovery of apricot trees from Verticillium wilt. Aust. J. Biol. Sci. 23:1027-1032.

Hawksworth, D. L., and Talboys, P. W. 1970. *Verticillium albo-atrum*. Descriptions of Pathogenic Fungi and Bacteria, Nos. 255 and 256. Commonwealth Mycological Institute, Kew, Surrey, England.

Howell, C. R. 1973. Pathogenicity and host-parasite relationships. Proc. Work Conf. Nat. Conf. Pathol. Res. Lab. (College Station, Tex., Aug. 30–Sept. 1, 1971) ARS-S-19:42-46.

McCain, A. H., Raabe, R. D., and Wilhelm, S. 1981. Plants resistant or susceptible to Verticillium wilt. Leafl. 2703. University of California, Division of Agriculture and Natural Resources, Oakland. 10 pp.

Parker, K. G. 1959. Verticillium hadromycosis of deciduous tree fruits. Plant Dis. Rep. Suppl. 255:39-61.

Wilhelm, S., and Ferguson, J. 1953. Soil fumigation against *Verticillium albo-atrum*. Phytopathology 43:593-596.

(Prepared by W. D. Gubler)

Phymatotrichum Root Rot

Phymatotrichum root rot is found only in the Western Hemisphere, where it is confined to the alkaline, calcareous soils of the southwestern United States and Mexico. In this region, it is a limiting factor in production of many fruit and nut crops (apples, pears, grapes, peaches, and pecans). The pathogen parasitizes over 2,000 species of taprooted plants.

Symptoms

Diseased trees show various degrees of debilitation. Leaves develop a slight yellowing or bronzing, followed by loss of turgidity. Susceptible trees die quickly; leaves may abscise but usually remain attached to the branches for some time, depending on the rate of plant death (Plate 61). If the pathogen is located on part of the root system, the aboveground symptoms are manifested in unilateral dying of the branches. Roots are usually extensively invaded by the pathogen at the time aboveground symptoms appear. Cortical tissue is killed, sloughs off easily, and is often covered by a network of visible strands of the fungus. During warm, rainy weather, light-colored spore mats are often seen on the soil surface.

Causal Organism

Phymatotrichum root rot is caused by *Phymatotrichopsis omnivora* (Duggar) Hennebert (syn. *Phymatotrichum omnivorum* (Shear) Duggar). The fungus rarely sporulates in vitro, and there is no conclusive evidence that a teleomorphic stage exists. *P. omnivora* forms two types of hyphae. Large-celled hyphae (20 μm in diameter), with septa 60–120 μm apart, are composed of multinucleate cells (up to 40 nuclei per cell) with right-angle branching similar to *Rhizoctonia solani* (Kühn). Small hyphae give rise to acicular hyphae that form characteristic branches on rhizomorph-like strands. The acicular, cruciform branches distinguish *P. omnivora* from other fungi (Plate 62).

Disease Cycle and Epidemiology

P. omnivora overseasons in the soil as sclerotia or mycelial strands on roots. When soil moisture and temperature are favorable, the pathogen resumes growth and colonization of roots. Mycelial strands radiate through the soil to encounter roots from adjacent trees. Once a root is infected, nutrients are translocated basipetally through strands, and new sclerotia form in the soil. The highest population levels of sclerotia are found 45–90 cm deep in the soil, but some sclerotia have been recovered from depths of 240 cm.

Control

Phymatotrichum root rot is extremely difficult to control in established orchards. Before an orchard is established, land should be planted to a herbaceous host (e.g., alfalfa or soybeans) to identify infested regions. Susceptible hosts should not be planted between rows in young orchards. Deep applications (50 gal/acre) of soil fumigants such as dichloropropene are effective eradicative control measures. There are no therapeutic systemic fungicides that will translocate basipetally and control Phymatotrichum root rot, and no rootstocks are resistant.

Selected References

Lyda, S. D. 1974. Studies on *Phymatotrichum omnivorum* and Phymatotrichum root rot. Pages 69-73 in: The Relation of Soil Microorganisms to Soilborne Plant Pathogens. G. C. Papavizas, ed. South. Coop. Ser. Bull. 183. Polytech. Inst. State Univ., Blacksburg, VA.

Lyda, S. D. 1978. Ecology of *Phymatotrichum omnivorum*. Annu. Rev. Phytopathol. 16:193-209.

Lyda, S. D., Robinson, G. D., and Lembright, H. W. 1967. Soil fumigation control of Phymatotrichum root rot in Nevada. Plant Dis. Rep. 51:331-333.

Rogers, C. H. 1942. Cotton root rot studies with special reference to sclerotia, cover crops, rotations, tillage, seeding rates, soil fungicides, and effects on seed quality. Tex. Agric. Exp. Stn. Bull. 614.

Streets, R. B., and Bloss, H. E. 1973. Phymatotrichum root rot. Monogr. 8. American Phytopathological Society, St. Paul, MN. 38 pp.

(Prepared by S. D. Lyda)

Rosellinia (Dematophora) Root Rot

Rosellinia root rot, also known as white root rot, is a serious disease of many deciduous fruit trees in most warm temperate and subtropical regions. The pathogen's host range includes 170 species in 63 genera and 30 families. In addition to stone fruit, the fungus attacks pome fruit, avocado, citrus, feijoa, olive, pistachio, and walnut, as well as bushberry, strawberry, loquat, fig, mango, and grape. In Japan, Brazil, and South Africa, stone fruits are severely affected, as are apples and pears. In the United States (California), the disease is found sporadically on pears and rarely on stone fruits. The anamorph of *Rosellinia* (*Dematophora*) has been reported on pome fruits in the eastern United States, but its occurrence has not been confirmed recently.

Symptoms

Infected trees may die very rapidly or decline for several years. Although trees become infected soon after planting, acute symptoms usually do not develop until they begin to fruit. Trees with chronic symptoms frequently bear more flower buds than leaf buds in the year before they die.

Aerial symptoms include premature yellowing of foliage and poor growth of one or all limbs and new shoots in spring. Fruit cease to grow and may shrivel or drop. Leaves are small and almost white, and some premature defoliation may occur. These symptoms are not distinctively different from those caused by attacks of other soilborne pathogens, such as *Armillaria mellea* (Vahl) P. Kumm. and *A. tabescens* (Scop.) Dennis, Orton & Hara, but *Rosellinia necatrix* produces a whitish, wefty strand or ribbon, as opposed to well-defined rhizomorphs.

The characteristic belowground symptom is the rotting of fibrous roots by a whitish, wefty to floccose layer of mycelium which then invades larger roots. The invasion extends through the cortex and cambium and then progresses upward into the trunk. Occasionally, the fungus ruptures the bark and extends into the adjacent soil. The well-defined, fan-shaped mats of mycelium with whitish, wefty strand may appear when the outer layer of dead bark is removed (Plate 63). Under suitable moisture and temperature conditions, white mycelium appears at ground level near the crown of the tree. Diseased trees can be removed easily from the soil, because roots deteriorate severely. Infected roots placed in a moist chamber quickly become overgrown with characteristic dense, white hyphae.

Causal Organism

Rosellinia necatrix Prill. (anamorph *Dematophora necatrix* R. Hartig), the causal organism, produces an ostiolate perithecium that is rarely found in nature. The perithecia are spheroidal, black, and erumpent and are produced on a mycelial crust. Asci are cylindrical (250–380 × 8–12 μm), and ascospores (30–50 × 5–8 μm) are one-celled, with a hyaline epispore. Characteristic pear-shaped swellings (geniculations) at the septa of the somatic hyphae are distinctive diagnostic characteristics. Coremia arising from sclerotia produce small, single-celled, colorless conidia (3–4.5 × 2–2.5 μm) on multiple-branched conidiophores. The fungus produces scattered black microsclerotia (98 × 130 μm), which tend to unite and form microsclerotial sheets.

Disease Cycle

Data on epidemiology are scarce from stone fruit-growing areas. *R. necatrix* can survive for several years on rotted roots in the ground and occasionally as mycelium or loose microsclerotia in the soil or adhering to root debris. Ascospores are produced so rarely as to be of little importance as inoculum. Viability of conidia often is low; infection takes place primarily from mycelium, which attacks the fine roots and apparently enters by direct penetration. The hyphae branch freely in the cortex, destroying it as they grow toward larger roots.

The fungus spreads by growth through the soil and infects roots of adjacent trees. Infested soil and diseased root pieces may be distributed by cultivation or on nursery stock. The fungus develops most rapidly in cool, moist soils.

Control

No entirely satisfactory control methods have been developed. Control measures consist largely of attempts to prevent spread of the fungus in the soil and to eliminate sources of inoculum. Soil fumigation with methyl bromide or chloropicrin can almost eliminate the pathogen in the soil, but not the mycelium in roots. Chloropicrin is applied to soil or injected during summer or fall. Covering the soil with vinyl sheet improves performance. In field experiments in Japan, benomyl, thiophanate methyl, and isoprothiolane prevented development of the fungus on roots during early stages of infection. These fungicides should be applied as a drench after exposing the crown of trees in an area 1 m in radius and 30 cm deep.

Selected References

Ochiai, M., and Hayashi, S. 1978. Studies on the chemical control of apple white root rot caused by *Rosellinia necatrix* (Hartig) Berlese. Bull. Fukushima Hortic. Exp. Stn. 8:25-44.

Sivanesan, A., and Holliday, P. 1972. *Rosellina necatrix*. Descriptions of Pathogenic Fungi and Bacteria, No 352. Commonwealth Mycological Institute, Kew, Surrey, England.

Sztejnberg, A. 1990. *Rosellinia* (*Dematophora*) root rot. Pages 46-47 in: Compendium of Apple and Pear Diseases. A. L. Jones and H. S. Aldwinckle, eds. American Phytopathological Society, St. Paul, MN.

Sztejnberg, A., and Madar, Z. 1980. Host range of *Dematophora necatrix*, the cause of white root rot disease in fruit trees. Plant Dis. 64:662-664.

Umemoto, S., and Murata, A. 1986. Attempt on the prompt and convenient control of white root rot of Japanese pear caused by *Rosellinia necatrix* Prilleux. Bull. Chiba-Ken Agric. Exp. Stn. 27:141-152.

(Prepared by Y. Terai)

Violet Root Rot

The violet root rot pathogen is endemic in Japan, Korea, and China and affects a broad range of fruit trees as well as vegetable crops such as sweet potato, asparagus, and carrots. In Japan, 113 species in 50 families of plants have been reported as hosts. The disease causes significant crop losses in Japan.

Symptoms

The aboveground parts of affected trees show symptoms similar to those of other diseases that affect roots (e.g., Rosellinia root rot): reduced shoot growth; small, light green leaves; dieback of twigs and shoots; and gradual decline of the whole tree. The affected root surface of an infected tree is covered with a purplish or dark brown network of mycelium. The mycelial growth is often thick and may extend up the roots to the trunk area near the soil surface. Fruiting bodies of the fungus are formed in autumn around the trunk at the soil line. The infected roots eventually rot and disintegrate.

Causal Organism

Violet root rot is caused by *Helicobasidium mompa* Tanaka, a basidiomycetous fungus in the family Auriculariaceae. The fungus grows at temperatures between 8 and 35°C (optimum 20–29°C). This fungus differs from *H. brebissonii* (Desmaz.) Donk (syn. *H. purpureum* Pat.), the violet root rot fungus that is widespread in Europe and the United States, which has a violet to purple fruiting body and bispored basidia. The fruiting body of *H. mompa* is resupinate, somewhat orbicular or oblong, often irregularly lobed, 5–10 cm in diameter and 2–4 mm thick, velvety, at first pliable then subcoriaceous, somewhat convex, incrustate, purplish brown, and at length albo-pruinose. The hymenium is white with curved basidia, which are one- to three-septate, and tetraspored basidiospores (10–12 × 5–7 μm), produced on elongated sterigmata, are ovoid, curved, and hyaline.

Disease Cycle and Epidemiology

The fungus from infested soil can infect roots or move from roots of infected trees to those of healthy trees. Mycelial strands extend along the roots to the soil surface and produce fruiting bodies near the trunk in autumn and basidiospores the following summer. The role of basidiospores in the spread of the pathogen is not clearly understood. Soils originating from volcanic ash that are rich in raw organic matter favor survival of this fungus.

Control

New plantings should be established in noninfested soils. In sites with infected trees, diseased trees and infested soils should be removed, and the soil should be treated with a soil sterilant such as chloropicrin before replanting. Curative methods that involve removing infected roots near the crown and treating the soil with captafol or amobam to prevent fungal spread have been successful on diseased apple trees and may also be effective on stone fruits. Rootstocks resistant to this fungus are not available.

Selected References

Araki, T. 1967. Soil conditions and the violet and white root rot diseases of fruit trees. (In Japanese, with English summary.) Bull. Nat. Inst. Agric. Sci. Ser. C. 21:1-109.

Harada, Y., and Segawa, K. 1990. Violet root rot. Pages 52-53 in: Compendium of Apple and Pear Diseases. A. L. Jones and H. S. Aldwinckle, eds. American Phytopathological Society, St. Paul, MN.

(Prepared by K. Sawamura)

Disease Complexes

Wood Decay

Wood decay of stone fruit trees is caused mostly by basidiomycetous fungi and occurs throughout the world wherever these trees are grown. Wood decay disorders represent distinct problems in orchard management and generally are observed in older trees after extensive fungal colonization of the wood has occurred (Plates 64–67). Orchard management problems associated with wood decay include broken limbs and toppled trees as well as tree dieback, decline, and death. Several reports of tree decline and surveys of wood decay fungi on *Prunus* species have been made in the states of California, South Carolina, and Pennsylvania (Table 2) and in Australia, Europe, and India. Pathogenicity studies have been done on one wood-rotting fungus on cherry in California.

Symptoms

Symptoms of wood decay are similar among species of *Prunus*. Two types of wood decay are recognized on living stone fruit trees—white and brown rots. They are usually distinctive in exposed wood, and the extensive damage they cause may result in structural failure of the wood. White rots, which are the most common, are generally moist, soft or spongy, and lighter in color than sound wood. All major structural components of the cell wall, including cellulose, hemicellulose, and lignin, are degraded. Strength properties of white-rotted wood are decreased only in the advanced stages of decay. In brown rot of wood, the decayed wood is brown, generally dry, and crumbly, with both longitudinal and transverse cracks. Most of the cell wall polysaccharides are degraded, but the lignin component is unchanged or slightly modified. Strength properties of brown-rotted wood are significantly reduced or lost entirely in a relatively short period after decay is initiated.

Macroscopic effects induced by these decays include limb breakage during fruit production, uprooted trees, trees broken at the soil line during windstorms or during mechanical harvesting, and decreased tree vigor and dieback. Symptoms caused by specific decay fungi may range from foliar symptoms (e.g., silver leaf) to tree death. Canker formation, tree gumming, and death can occur when decay fungi such as species of *Armillaria* invade cambial tissue (see Armillaria and Clitocybe root and crown rots).

Causal Organisms

Fungi that cause extensive wood decay of stone fruit trees are classified primarily in the class Hymenomycetes of the subdivision Basidiomycotina (Table 2). These fungi usually produce a fruiting body called a basidiocarp or basidiome and form specialized cells called basidia, where sexual reproduction occurs. Basidia are arranged in a fertile layer called the hymenium and produce sexual spores called basidiospores. The hymenial tissue can be arranged in numerous ways,

TABLE 2. Common Species of Decay Fungi, Type of Decay, and Location on Stone Fruit Trees in North America

Genera and Species of Hymenomycetes	Type of Decay[a]	Location on Tree[b]	Associated Injury[c]
Antrodia albida (Fr.:Fr.) Donk	Brown	S	P
Armillaria mellea (Vahl:Fr.) P. Kumm.	White	RT	NW
Fomitopsis cajanderi (P. Karst.) Kotlaba & Pouzar	Brown	TS	PMt
F. meliae (Underw.) R.L. Gilbertson & Ryvarden	Brown	TS	PMt
Ganoderma brownii (Murrill) R.L. Gilbertson	White	RT	PMt
G. lucidum (Curtis:Fr.) P. Karst.	White	RT	PMt
Irpex lacteus (Fr.:Fr.) Fr.	White	TS	P
Laetiporus sulphureus (Bull.:Fr.) Murrill	Brown	RTS	PMt
Oxyporus latemarginatus (Durieu & Mont.) Donk	White	RT	MrMt
Phellinus gilvus (Schwein:Fr.) Pat.	White	TS	PMt
P. robustus (P. Karst.) Bourd. & Galzin	White	TS	PMt
Schizophyllum commune Fr.:Fr.	White	S	CstSP
Schizopora paradoxa (Schrad.:Fr.) Donk	White	S	P
Stereum hirsutum (Willd.:Fr.) S.F. Gray	White	S	PMt
Trametes hirsuta (Wulfen:Fr.) Quel.	White	TS	PMt
T. versicolor (L.:Fr.) Pilát	White	TS	PMt
Trichaptum biforme (Fr.) Ryvarden	White	TS	PMt

[a] Type of decay: white = white rot, and brown = brown rot of wood.
[b] Location on tree: R = root, S = scaffold branch, and T = trunk.
[c] Associated injuries that decay fungi commonly colonize: P = pruning wounds; NW = wounds not necessary for infection; Mt = mechanical injuries on trunks; Mr = mechanical injuries on roots; Cst = chemical injuries on scaffold branches and trunks (e.g., oil damage); and S = sunburn wounds.

depending on the basidiocarp of the fungus. This tissue can be found lining the inner walls of tubes, comprising the surface of lamellae or gills, or composing the surface of a smooth or flat (i.e., resupinate, effused-reflexed) fruiting body, depending on the species of the fungus.

Basidiospores of wood decay fungi are discharged forcibly from the basidia and are usually disseminated by air currents. Once deposited on a suitable substrate, the basidiospores may germinate when conditions are favorable. In heterothallic species, basidiospores germinate to form homokaryotic mycelia that fuse with compatible mating types to establish a heterokaryon. In some species of *Armillaria,* however, nuclei in a dikaryotic mycelium commonly fuse to form a diploid mycelium. No hyphal fusion is required in homothallic species. Mycelial growth of the fungus proceeds with colonization and utilization of wood as a food source. Basidiocarps develop after extensive substrate colonization under suitable environmental and physiological conditions. Asexual spores are not commonly produced.

Disease Cycle and Epidemiology

Wood decay fungi enter trees primarily through wounds that expose secondary xylem of sapwood or heartwood. In stone fruit trees, injuries from pruning, mechanical harvesting, sunburning, branch supports, and cultivating equipment, as well as from other pathogens (e.g., *Nectria cinnabarina* Fr.), can expose susceptible wood. In the field, decay fungi commonly function as secondary invaders of tree wood, following a succession of microorganisms including bacteria, yeasts, and other fungi. Wood decay fungi, however, are capable of decaying wood in the absence of other organisms. Some decay fungi, such as species of *Armillaria* and *Ganoderma*, can spread through natural root grafts and thus function as primary invaders. Extensive decay of heartwood, and in some cases sapwood, occurs once the decay fungus is established in the tree. Some species of decay fungi are restricted to the roots and lower trunk of the tree; others are commonly found on scaffold and secondary branches, and still others can be found on all parts of the tree (roots, trunk, and branches). Furthermore, decay in a single tree may or may not be limited to one species of fungus.

Control

Control of wood rot disorders involves cultural practices that promote tree growth and vigor, while minimizing injuries that expose wood. Orchard management begins with proper site and cultivar selection, followed by appropriate programs for irrigation and fertilization. Pruning should be done when branches are small to enhance wound healing, since large wounds are often points of entry for decay fungi. In addition, pruning cuts should be made adjacent to but not into the supporting branch to allow for proper callus formation and natural wound healing. Therefore, stub cuts and horizontal cut surfaces that inhibit wound healing and allow for water accumulation should be avoided.

Sanitation may also help reduce infection. Fruiting bodies and dead branches should be removed and destroyed to prevent the formation of new inoculum. Ideally, pruning should be done when inoculum is sparse or absent. However, numerous species of wood decay fungi are recognized on stone fruit trees, and they have different peak sporulation periods depending on the species and the environment, as influenced by geographic location. Investigations of sporulation periods of decay fungi in specific geographic regions may be beneficial when developing forecasts of low-risk periods for pruning.

Tree paints have been used to protect wounds, but their usefulness is uncertain. Wound protection depends on complete coverage and maintenance of an intact layer of protectants. Cracking of the sealant may expose unprotected wood and allow the establishment of decay fungi. Systemic fungicides applied to exposed wood wounds may provide extended protection. Several of these materials, including triadimenol, have provided wound protection. Other methods used to control wood decay fungi include biological control and host resistance.

Selected References

Adaskaveg, J. E., and Ogawa, J. M. 1990. Wood decay pathology of fruit and nut trees in California. Plant Dis. 74:341-352.

Ogawa, J. M., and English, H. 1991. Diseases of Temperate Zone Tree Fruit and Nut Crops. Publ. 3345. University of California Division of Agriculture and Natural Resources, Oakland. 461 pp.

(Prepared by J. E. Adaskaveg and R. L. Gilbertson)

Peach Tree Short Life

Peach tree short life (PTSL) is a disease complex characterized by a sudden wilt and collapse of newly emerged flowers and leaves of peach trees and death of branches or the entire aerial portion of the tree. PTSL also affects other stone fruits, including nectarine, apricot, and plum. Peach tree decline and peach tree replant problem are other names used for this disease, but these names also may refer to any problem that diminishes normal tree growth and productivity, especially after peach orchards are replanted. The cause of PTSL is more specific and better understood than some of the more general orchard decline and replant problems.

In the southeastern United States, several million peach trees have been killed by PTSL during the past 30 years. In sites prone to PTSL, tree mortality may exceed 50% in one year. Commonly, apparently healthy trees may thrive immediately adjacent to dead or dying trees.

There are many similarities between PTSL and the bacterial canker complex in California. Cold injury, which kills many trees affected by PTSL, is not a major contributing factor in California, but most other factors and their interactions are similar. Ring nematodes are important initiators of both syndromes.

Symptoms

Trees in their third to sixth year after planting are most susceptible, although older or younger trees can be affected. In the typical sequence, apparently normal tree growth occurs for the first two to five years after planting. Wilt symptoms appear suddenly in the spring, and tree death often follows within a few weeks. Some acutely affected trees die during late winter before bloom. During warm spring weather, bark on the trunk, crotch, and scaffold limbs develops a reddish, wet appearance, often accompanied by droplets of yellow or orange exudate (Plate 68). A "sour-sap" odor often is associated with tissues with this appearance. Further examination of the trunk and scaffold limbs may reveal cracks in the outer bark that usually extend inward to the xylem. The inner bark and the cambial area are discolored, but the discoloration extends only to the soil line, thus differentiating PTSL from collar and root rots. In general, primary roots appear healthy, but feeder roots may be sparse, discolored, or necrotic. Primary symptoms can become masked quickly by saprophytes that colonize the damaged and necrotic tissues. It is common for new shoots to grow from the base of the trunk by mid- or late summer.

Causal Factors

PTSL is a complex disease, with numerous biotic and abiotic factors interacting. Causal agents can be described as those directly associated with tree death and those that predispose trees to the direct factors. Direct factors are freeze injury, bacterial canker (caused by *Pseudomonas syringae* pv. *syringae*), and perhaps Leucostoma canker (caused by *Leucostoma persoonii* Höhn.); predisposing factors include the ring nematode (*Criconemella xenoplax* (Raski) Luc & Raski), rootstocks, time of pruning, root injury caused by cultural practices, and physical characteristics of the orchard site, such as pH and soil structure.

Continuous presence of moderate to high densities of *C. xenoplax* (>50/100 cm^3 of soil) has preceded and often accompanied PTSL symptoms. In orchards where preplant and postplant nematicides have been used to suppress *C. xenoplax*, tree losses from PTSL are negligible or greatly reduced. Scions budded onto rootstocks such as Nemaguard, Yunnan, Shalil, and S-37, which are resistant to root-knot nematodes (*Meloidogyne* spp.), are especially susceptible to PTSL. Trees pruned in late fall or early winter (November to January in the southeastern United States) are more likely to succumb to PTSL than are trees pruned later (after 1 February). The predisposing effect of untimely pruning is most pronounced in trees three to six years old, but the mechanism of predisposition is unknown. Cultural practices that damage tree roots, such as excessively deep cultivation, also may increase susceptibility to PTSL. Trees grown in acidic soils are more susceptible than trees grown in soils where the soil pH is maintained in the range of 6.0–6.5. Soil pH may affect root growth, tree vigor, soil and root microflora, nematodes, and susceptibility to *P. syringae* pv. *syringae,* or it may cause aluminum to become available for root uptake. Compacted soil or a hardpan also may inhibit tree growth and enhance susceptibility.

Disease Cycle and Epidemiology

PTSL is most common and severe where trees are planted in light, sandy soil, especially in locations where peach trees have been grown previously. In the southeastern United States, freeze injury is the most common factor directly responsible for tree death in the PTSL complex. Brown discoloration of the inner bark and cambial area follows freeze injury, and the outer bark often cracks and is easily separated from the wood. Although the inner bark and cambium may be severely damaged, flower buds are seldom killed at this time. Severe tree losses often occur when winter (January and February) temperatures reach 22°C or higher for several days, then drop rapidly to −10°C or less.

P. syringae pv. *syringae* (the cause of bacterial canker) is associated with bud and twig death, but in trees prone to PTSL, cankers extend from the dead buds into larger branches and from there into trunks of trees. Death of large portions or entire trees may follow. Bacterial canker is observed most often when winter temperatures are mild and rains are frequent. *Leucostoma* spp. commonly colonize dead twigs, wounds, and branches. The role of *Leucostoma* in aggravating or extending the area of injured tissue is uncertain, but perennial cankers usually do not develop in the southern United States. In late spring, the bark of trunks, crotches, and scaffold limbs injured by freezing temperatures or by *P. syringae* pv. *syringae* may be extensively colonized by *Leucostoma* and covered with pycnidia, which sporulate profusely when moist.

Control

The direct causes of mortality, i.e., freeze injury, bacterial canker, or *Leucostoma* infections, are difficult to control. The predisposing agents, which increase the susceptibility of trees to the direct factors, are more easily managed. The following 10-point management scheme was developed in the early 1970s and recommended for use in the southern United States:

1. Before planting, apply lime to adjust soil pH in the top 20 cm to 6.0–6.5.
2. Subsoil during site preparation to break up the hardpan, thereby improving water infiltration, drainage, root growth, nutrient uptake, and diffusion of nematicides.
3. In sandy soils where peach trees have been grown previously and in other soils where ring and root-knot nematodes are a problem, fumigate the soil before plant-

ing trees.

4. Plant trees that have been grown in fumigated soil or in soil free of parasitic nematodes and other diseases.
5. Plant trees propagated on Lovell or Halford rootstocks (both are very susceptible to root-knot nematodes; thus, preplant fumigation often is essential).
6. Apply nutrients and lime as needed based on soil tests, foliar analysis, and local recommendations.
7. Prune as late as possible, never before 1 January and preferably after 1 February. If earlier pruning is unavoidable, prune older trees first. Early pruning is especially hazardous for trees grown on locations where peaches were previously grown. Discontinue summer pruning (including topping and hedging) by 15 September.
8. Use recommended herbicides for weed management. Mechanical cultivation, if used, should be shallow to avoid root injury.
9. In sites where preplant fumigation was necessary, use a postplant nematicide if ring nematode populations increase. Assay soil for nematodes annually.
10. Promptly remove from the orchard and destroy all dead and dying trees.

Implementation of this 10-point program in research plots and growers' orchards has substantially reduced losses to PTSL. Currently, nematode control is the most difficult practice to achieve for effective, cost-efficient management of PTSL. Recent progress in biological control and development of nematode-tolerant rootstocks may lead to more effective management of this disease complex.

Selected References

Lownsbery, B. F., English, H., Noel, G. R., and Schick, F. J. 1977. Influence of Nemaguard and Lovell rootstocks and *Criconemoides xenoplax* on bacterial canker of peach. J. Nematol. 9:221-224.

Nyczepir, A. P., Zehr, E. I., Lewis, S. A., and Harshman, D. C. 1983. Short life of peach trees induced by *Criconemella xenoplax*. Plant Dis. 67:507-508.

Prince, V. E., and Horton, B. D. 1972. Influence of pruning at various dates on peach tree mortality. J. Am. Soc. Hortic. Sci. 97:303-305.

Ritchie, D. F. 1989. Improved peach tree longevity with use of fenamiphos in peach tree short-life locations. Plant Dis. 73:160-163.

Ritchie, D. F., and Clayton, C. N. 1982. Peach tree short life: A complex of interacting factors. Plant Dis. 65:462-469.

Wehunt, E. J., Horton, B. D., and Prince, V. E. 1980. Effects of nematicides, lime, and herbicide on peach tree short life in Georgia. J. Nematol. 12:183-189.

Zehr, E. I., Miller, R. W., and Smith, F. H. 1976. Soil fumigation and peach rootstocks for protection against peach tree short life. Phytopathology 66:689-694.

(Prepared by D. F. Ritchie and E. I. Zehr)

Replant Disorders

Stone fruit trees planted into soil recently cropped to other stone fruits often grow poorly for one to several years and lag behind healthy trees in growth and onset of fruit production. Such stone fruit replant disorders (SFRDs) lack obvious aboveground symptoms other than reduced growth and increased mortality during their first three to five years of growth. SFRD severity usually is greatest where a fruit crop follows itself, but both specific (e.g., peach after peach) and nonspecific (e.g., cherry after plum) replant disorders are known. Stone fruits also may grow poorly in sites that follow vegetables or alfalfa, where the same nonspecific causes are present. Regardless of the type of replant disorder involved, the primary, but often overlooked, cost is delayed and reduced fruit production over the life of the tree.

Replant disorders occur in stone fruit production areas worldwide, including those in Africa, Europe, Greece, Israel, Japan, New Zealand, and North America.

Symptoms

Affected trees typically exhibit reduced growth during their first few years after planting. The trees produce fewer and smaller leaves and shoots, which often have a lighter color than does normal growth. Shoot growth slows prematurely or stops completely in early summer. Root systems of such trees have few fine feeder roots, often with darkened or discolored necrotic lesions. There is no bark cracking or cambial/inner bark odor and discoloration such as that found in the lower branches and trunks of trees affected with peach tree short life. Increased mortality of 7–50% has been reported in replant sites in the northeastern United States and Ontario, Canada, and is accentuated by severe winters. In the arid southwestern United States, mortality rarely exceeds 10% per year and declines after the third year; drought stress also can increase replant mortality there. In many orchards, annual tree growth improves after one to several years, but affected trees never equal nonaffected trees in trunk size or cumulative yield over their expected life.

Causal Organisms

Both abiotic and biotic causes have been linked to specific and nonspecific SFRDs, and often an interaction of these factors appears to be involved. Abiotic factors include poor soil structure and drainage, strongly acidic or basic soil pH, toxicity of heavy metals (especially arsenic from arsenical insecticides), and unbalanced soil nutrition. In addition, soilborne organisms can break down prunasin present in old peach roots and release cyanide in sufficient concentrations to injure or kill the new peach feeder roots. Wet soils accentuate this cyanogenesis.

Several species of fungi and nematodes have been implicated in SFRDs. Fungi associated with stone fruit replant problems include *Thielaviopsis basicola* (Berk. & Broome) Ferraris, *Pythium* spp., *Cylindrocarpon* spp., *Phytophthora* spp., *Fusarium* spp., *Armillaria mellea* (Vahl.:Fr.) P. Kumm., *Clitocybe tabescens* Bres., and *Peniophora sacrata* G. Cunn. Nematodes usually are linked with nonspecific SFRDs. Those known to cause severe root damage and stunting of stone fruit trees include root-lesion nematodes (*Pratylenchus* spp.), ring nematodes (*Criconemella* spp.), root-knot nematodes (*Meloidogyne* spp.), and sometimes dagger nematodes (*Xiphinema* spp.). Nematodes tend to be a problem in light-textured, well-drained soils, while fungi are more commonly associated with heavier, poorly drained soils; in Europe, *Pratylenchus* reduces growth of replant cherry in sandy soils, whereas *Thielaviopsis* is a major problem in clay soils. A lack of vesicular-arbuscular mycorrhizal fungi also can cause replant problems for stone fruits.

Disease Cycle and Epidemiology

SFRDs develop in orchard soils after a few years of continuous stone fruit growth, but symptoms may not be evident until stone fruit are subsequently replanted. Severity in the subsequent crop is related to the number of years a site was cropped to stone fruit. SFRD effects may persist from two to six (or more) years where fallow or rotation crops are used for mitigation; this variation is due to the different biotic and/or abiotic factors at a given site and their ease of amelioration.

Control

The diversity of potential agents makes it necessary *either* to employ very broad-spectrum control measures (e.g., soil fumigation and/or general corrective cultural adjustments such

1. Blossom and twig blight of apricot caused by *Monilinia laxa.*

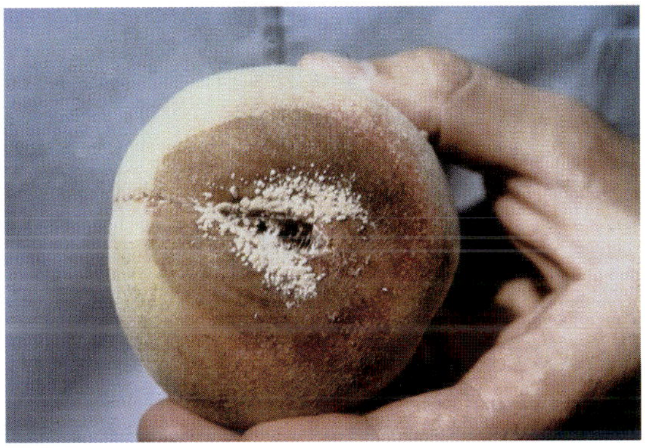

2. Decay of peach fruit caused by *Monilinia fructicola.*

3. Quiescent infections caused by *Monilinia fructicola* shown as small flecks on prune fruit. (Reprinted, by permission, from Ogawa and English, 1991)

4. Apothecia of *Monilinia fructicola* on a peach fruit mummy removed from soil and shown next to peach blossoms and mummy that remained attached to tree. (Reprinted, by permission, from Ogawa and English, 1991)

5. Typical morphologies of cultures of *Monilinia* on 2% potato-dextrose agar (*M. laxa* upper left, *M. fructicola* upper right, and *M. fructigena* bottom center).

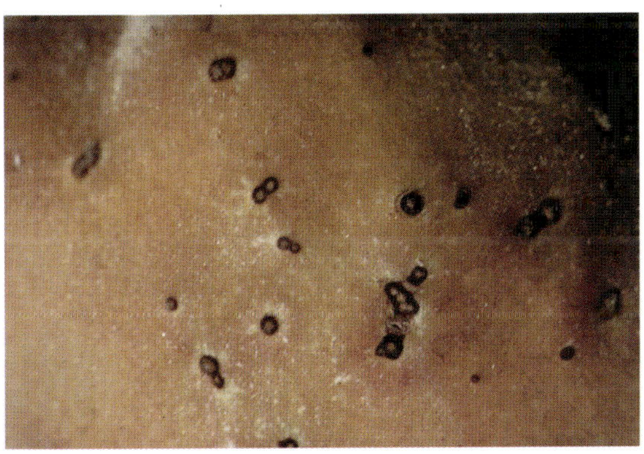

6. Peach fruit with shot-hole lesions.

7. Peach buds infected with shot-hole fungus. Note gumming.

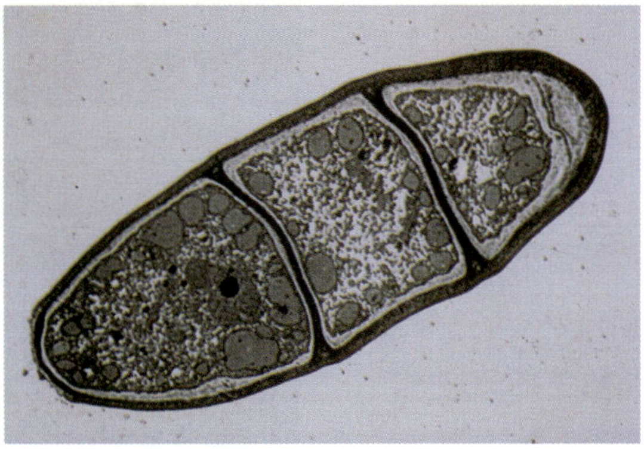

8. Electron micrograph of *Wilsonomyces carpophilus* conidium with double-walled construction.

9. Typical dark lesion of scab with characteristic halo on mature peach fruit. (Courtesy D. F. Ritchie)

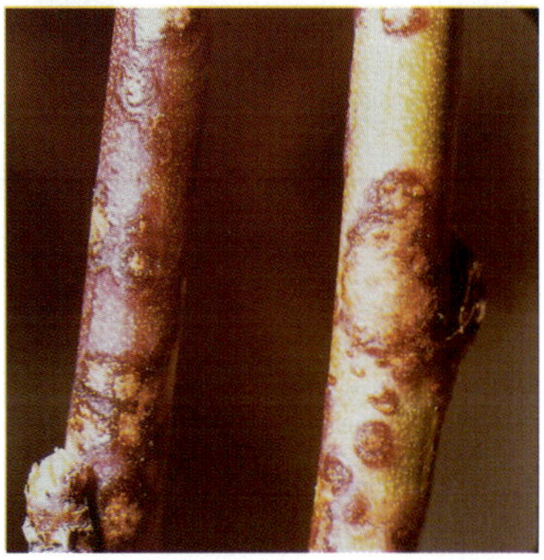

10. Discrete scab lesions on peach twigs. Note reddish margins where conidia are produced. (Courtesy E. I. Zehr)

11. Mature peach leaf with scab lesion on upper surface (top) and on lower surface (bottom). (Courtesy D. F. Ritchie)

12. Severely infected leaf with white powdery mass of mycelia and spores of *Sphaerotheca pannosa.*

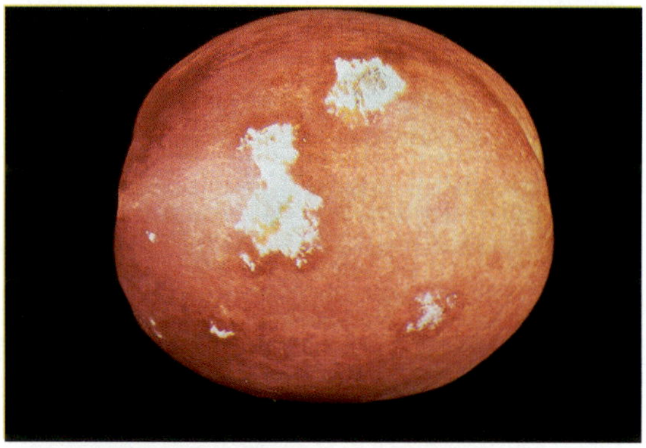

13. Nectarine fruit infected with powdery mildew. Note felt-like mat of mycelial growth.

14. Hypersensitive reaction of green apricot fruit to *Sphaerotheca pannosa* as indicated by brown necrotic tissue.

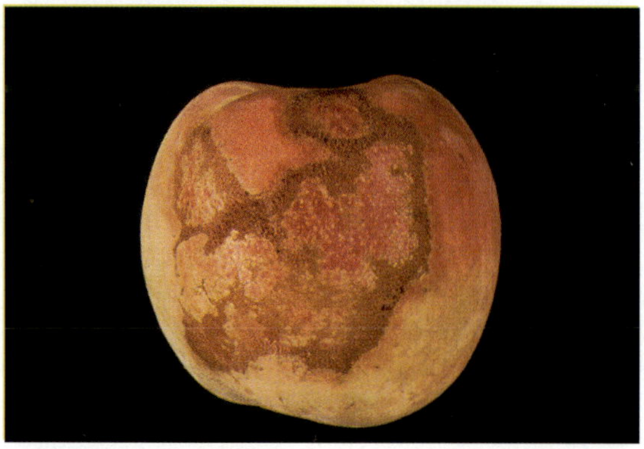

15. Rusty spot symptoms caused by *Podosphaera leucotricha* on peach fruit. (Courtesy B. T. Manji)

16. Invasion of *Sclerotinia sclerotiorum* from blighted blossoms into immature apricot fruit. (Reprinted, by permission, from Ogawa and English, 1991)

17. Immature prune fruit with healing scars (left) and scar with blossom parts infected by *Botrytis cinerea* (right). (Courtesy J. M. Ogawa)

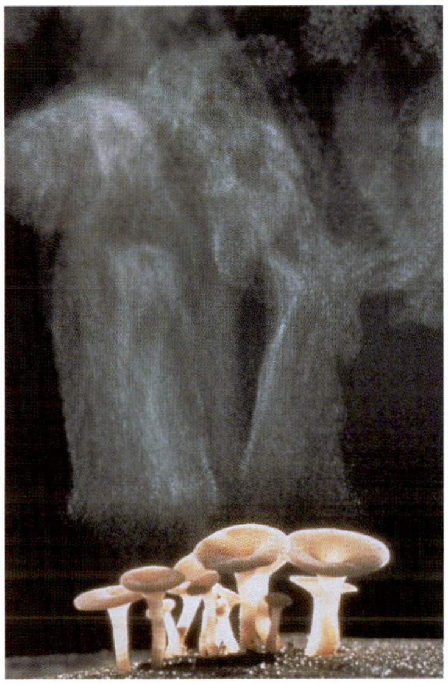

18. Ascospore ejection from apothecia of *Sclerotinia sclerotiorum*. (Copyright by the Regents of the University of California, Division of Agricultural and Natural Resources)

19. Rhizopus rot (left) compared with Gilbertella rot (right) on peach fruit. (Reprinted, by permission, from Ogawa and English, 1991)

20. Distinctive aerial mycelia and sporangiophores of *Mucor piriformis* on peach fruit.

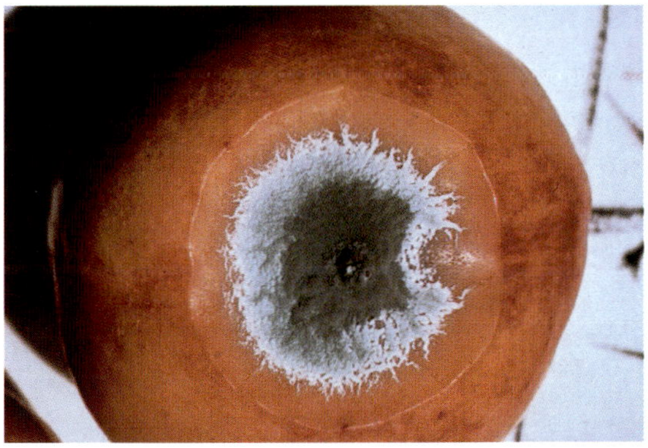

21. Penicillium rot on peach. (Reprinted, by permission, from Ogawa and English, 1991)

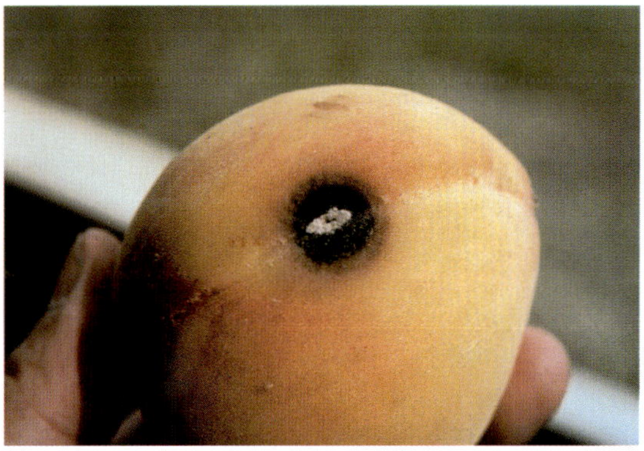

22. Alternaria rot on split stylar end of an apricot.

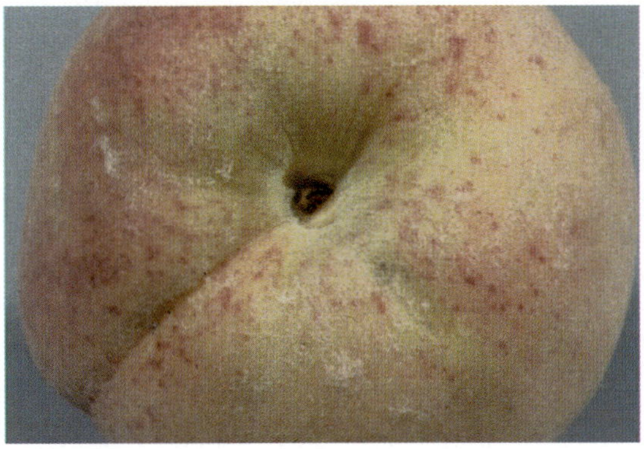

23. Red spot blemish on upper surface of peach fruit.

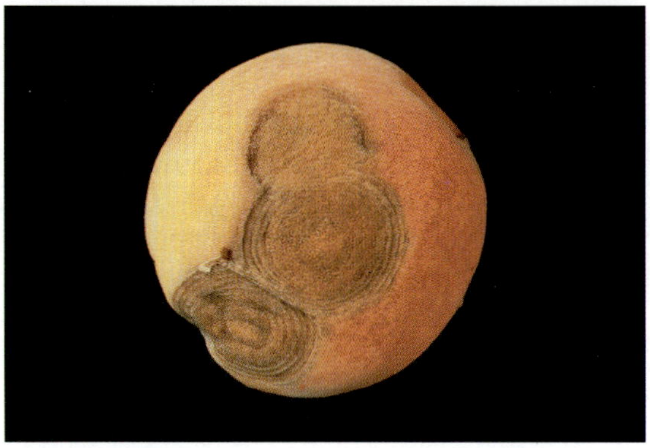

24. Anthracnose (*Colletotrichum gloeosporioides*) lesions with concentric zones and orange acervuli (spore masses) on Red Haven peach fruit. (Courtesy E. I. Zehr)

25. Large deformed fruit symptomatic of plum pocket (*Taphrina communis*) on left, normal plum fruit on right. (Courtesy T. Sutton)

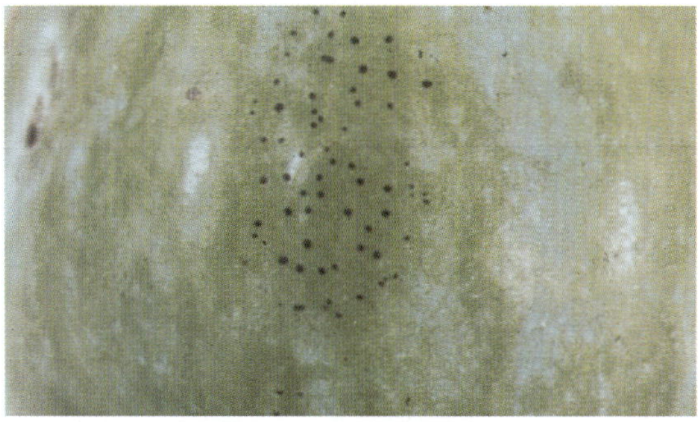

26. Flyspeck lesion, caused by *Zygophiala jamaicensis*, expressed as sharply distinct, black, shiny dots on plum. (Courtesy H. Nasu)

27. Acervuli and masses of conidia of *Phloeosporella padi* on the lower surface (left) and green lesions on upper, chlorotic leaf surface (right) of sour cherry infected with cherry leaf spot.

28. Stems of sour cherry fruit with lesions that contain white conidia and sporulation of the cherry leaf spot fungus (*Phloeosporella padi*). Note single lesion on the fruit (right).

29. Apothecia of the leaf spot fungus *Blumeriella* on overwintered sweet cherry leaves. (Courtesy T. B. Sutton)

30. Developing peach leaves with puckered areas caused by *Taphrina deformans*.

31. Reddish, raised areas on peach fruit caused by the peach leaf curl fungus, *Taphrina deformans*. (Courtesy P. F. Bertrand)

32. Witches'-broom symptom on shoot tip of sweet cherry infected with *Taphrina cerasi.*

33. Symptoms of rust caused by *Tranzschelia discolor* on upper (top) and lower (bottom) surfaces of peach leaves. (Courtesy J. M. Ogawa)

34. Monilia leaf blight caused by *Monilinia kusanoi.* Note white, powdery masses of conidia on the midrib and veins of cherry leaves.

35. Apothecia of *Monilinia kusanoi* produced on mummified cherry fruit.

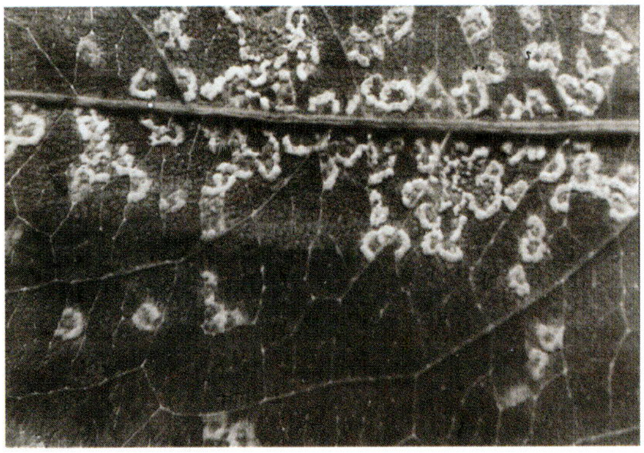

36. Symptoms of cup-shaped, white aecia of *Leucotelium pruni-persicae* on underside of peach leaf infected with Leucotelium white rust. (Courtesy H. Kitajima)

37. Shot-hole symptoms of zonate leaf spot on Japanese plum.

38. Cercospora leaf spot symptoms caused by *Cercospora circumscissa* on sour cherry leaves.

39. Leucostoma cankers caused by *Leucostoma persoonii* initiated around nodes of peach shoot. (Courtesy A. L. Jones)

40. Peach tree with Leucostoma cankers on scaffold limbs.

41. Peach tree with typical silver leaf symptoms: silver gray coloration of leaves and necrotic shoot tips.

42. Bracketlike fruiting bodies (basidiocarps) of *Chondrostereum purpureum* on the surface of the trunk of a dead peach tree.

43. Flagging and withering of peach leaves distal to twig cankers of constriction canker caused by *Phomopsis amygdali.*

44. Tendrils (cirri) of conidia exuding from pycnidia of *Phomopsis amygdali* on peach.

45. An orchard with severe Eutypa dieback in California's Suisun Valley. (Reprinted, by permission, from Ogawa and English, 1991)

46. Stroma of *Eutypa lata* sliced to show the perithecial locules. (Courtesy W. J. Moller)

47. Peach branch with blisters caused by *Botryosphaeria dothidea*. Note resin exudate.

48. Three-year-old trunk of peach with sunken necrotic lesions under bark.

49. Black knot on twig of Stanley prune.

50. Wine-colored inner bark of a Ceratocystis canker on a scaffold branch of French prune. (Reprinted, by permission, from Ogawa and English, 1991)

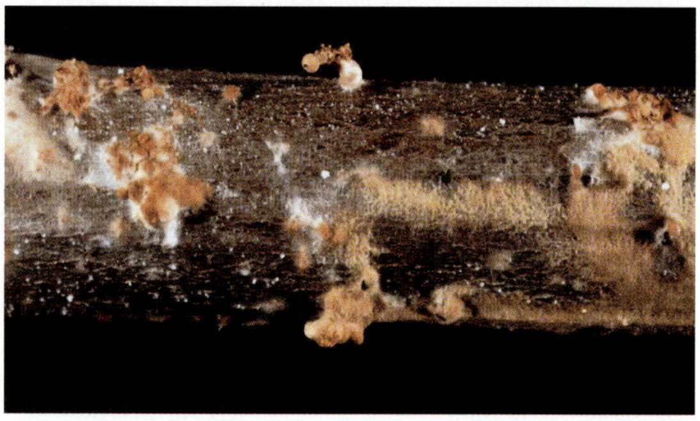

51. Amber-colored cirri (spore tendrils) exuding from the pycnidia of *Rhodosticta quercina* formed on plum branch.

52. Collapse of sour cherry tree with leaves attached caused by infection of the roots by an *Armillaria* sp.

53. White mycelia at cambial area (left) and shallow cut with rhizomorphs on surface (right) of an *Armillaria* sp. on the trunk of sour cherry trees.

54. Mushrooms of *Armillaria ostoyae* at base of a sour cherry tree. Note annulus attached to stipe.

55. Mushrooms of *Armillaria tabescens.* Note the lack of an annulus attached to the stipe. (Courtesy T. G. Beckman)

56. Cherry orchard with healthy cherry tree on mazzard rootstock (left) and collapsed tree with *Phytophthora* infection on mahaleb rootstock (right). (Courtesy J. M. Ogawa)

57. Peach tree with outer bark removed to show brown discoloration symptomatic of Phytophthora crown rot.

58. Initial symptom of Verticillium wilt of cherries expressed as withering of leaves on spurs low in the canopy.

59. Scaffold branch of peach tree with one-sided symptom of Verticillium wilt. (Courtesy W. J. Moller)

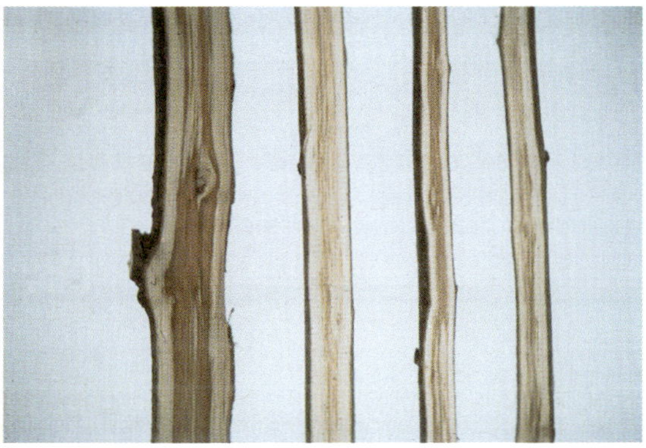

60. Characteristic darkening of xylem in plum infected with *Verticillium dahliae*.

61. Sudden death of peach tree infected with *Phymatotrichopsis omnivora* resulting in nonabscised leaves.

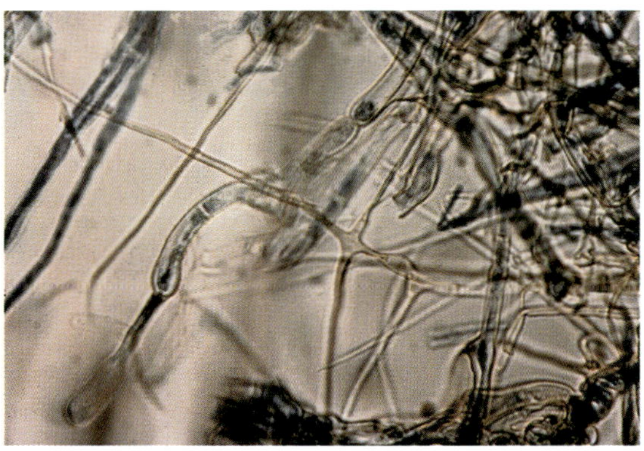

62. A typical cruciform, acicular hyphal branch of *Phymatotrichopsis omnivora*.

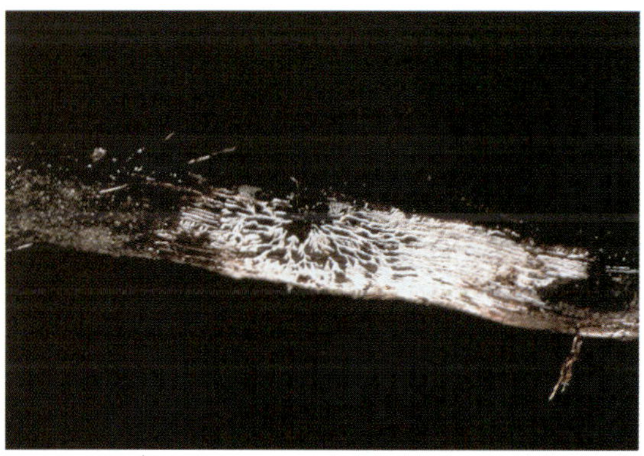

63. Infected roots with well-defined mycelial mat of *Rosellinia necatrix* under the bark.

64. Basidiocarps of *Trametes versicolor* developing on peach. The fungus causes a white rot of the wood.

65. Basidiocarp of *Ganoderma lucidum* developing on lower trunk of peach. The fungus causes a white rot of wood.

66. Basidiocarp of *Fomitopsis cajanderi* developing next to a pruning wound on a prune tree. The fungus causes a brown rot of the wood.

67. Resupinate basidiocarp of *Oxyporus latemarginatus* on branch stub. The fungus causes a white rot of wood.

68. Lower scaffold limbs of peach tree with gum and exudate characteristic of peach tree short life.

69. Leafy mistletoe (*Phoradendron tomentosum* subsp. *macrophyllum*) on mahaleb cherry.

70. Bacterial canker and dead buds on sour cherry. (Courtesy A. L. Jones)

71. Necrotic leaf spots along the margin of sweet cherry leaves caused by *Pseudomonas syringae* pv. *syringae*. (Courtesy A. L. Jones)

72. Necrotic spots with shot-holing on yellowing leaves (left) and fruit infection (right) on sour cherry caused by *Pseudomonas syringae* pv. *syringae*. (Courtesy A. L. Jones)

73. Blossom blast on prune associated with *Pseudomonas syringae* pv. *syringae* infection and freezing temperatures. (Courtesy A. L. Jones)

74. Bacterial decline. Note characteristic brown necrosis of bark at the tree crotch which fades progressively downward to give an undetectable margin with healthy bark.

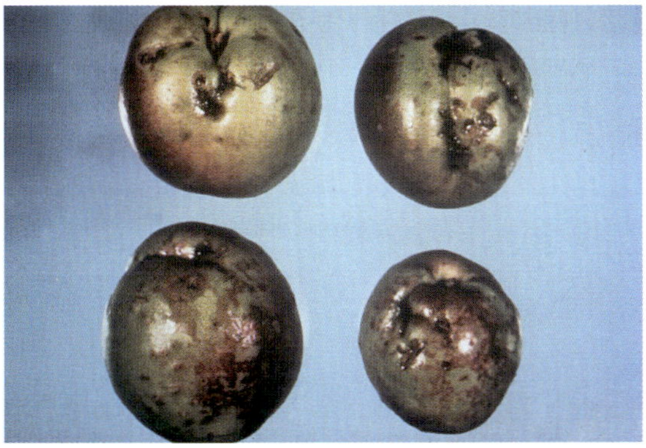

75. Lesions on immature nectarine fruit caused by *Pseudomonas syringae* pv. *persicae*.

76. Bacterial spot on peach leaves with developing leaf chlorosis.

77. Spring cankers caused by *Xanthomonas campestris* pv. *pruni* with cracked bark and shallow cankers on previous summer's twig growth.

78. Black tip on peach shoot caused by *Xanthomonas campestris* pv. *pruni*.

79. Bacterial spot on peach fruit.

80. Bacterial spot on nectarine fruit with cracking and gum exudation.

81. Crown gall on peach. (Courtesy D. F. Ritchie)

82. Peach tree with phony peach disease. Note compact growth habit and flat canopy. (Courtesy P. C. Andersen)

83. Marginal necrosis and leaf roll symptomatic of plum leaf scald. (Courtesy A. J. Latham)

84. Flower symptoms of the cherry blossom anomaly MLO. Note green coloration in the normally all-white petals.

85. Symptoms of pointed fruit diagnostic of cherry blossom anomaly MLO (healthy fruit at left).

86. Internal red color along the fruit suture of fruit infected with the peach red suture MLO. (Courtesy R. W. Miller)

87. Compact cluster of peach leaves caused by the peach rosette MLO. (Courtesy J. A. Foster)

88. Proliferation of upright, slender shoots with small chlorotic leaves on a scaffold branch of a tree infected with peach yellows MLO. (Courtesy J. K. Uyemoto)

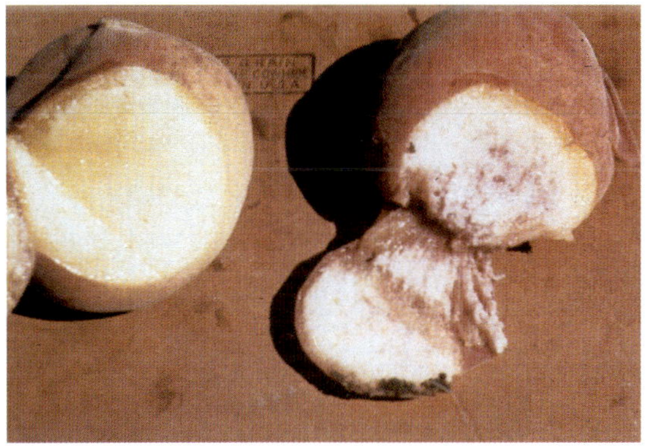

89. Reddish streaks in flesh of fruit with peach yellows MLO (healthy fruit at left). (Courtesy A. L. Jones)

90. Terminal leaf symptoms (upward-folded leaves and reddish color) on Bing cherry scions on mahaleb rootstock; scion infected with X-disease MLO.

91. Dark pits at the graft union of Bing cherry scion infected with X-disease MLO on mahaleb rootstock (healthy stock at right).

92. Small, triangular-shaped fruit with short pedicels from Bing cherry trees infected with X-disease MLO (healthy fruit at left).

93. Peach leaves with chlorotic and necrotic lesions caused by the X-disease MLO.

94. Peach yellow leaf roll MLO symptoms (enlarged midrib and lateral veins).

95. Severe leaf roll and chlorosis on new terminal growth of apricot infected by apricot chlorotic leaf roll MLO (healthy leaves at right).

96. Rudimentary buds and short internodes produced in the fall on shoots of apricot infected with chlorotic leaf roll MLO (healthy shoot at right).

97. Leaf reddening and rolling of Japanese plum infected with plum leptonecrosis MLO (healthy leaves at right).

98. Phloem necrosis in Japanese plum infected with plum leptonecrosis MLO.

99. Symptoms of leaf rolling and curling in peach (cv. South Haven) infected by apricot chlorotic leaf roll MLO (healthy leaves at left).

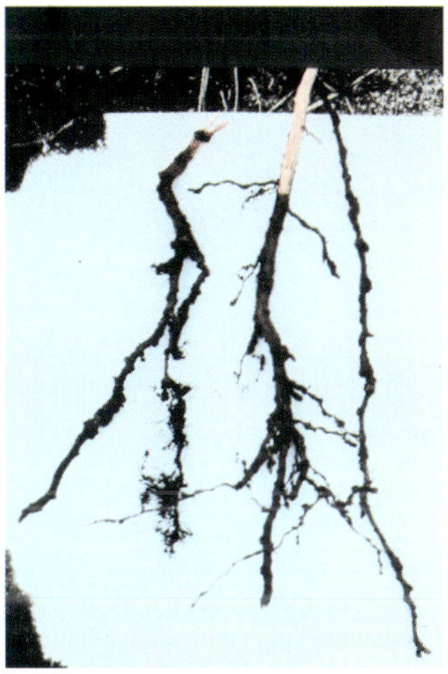

100. Root-knot nematode galls on Lovell peach roots. (Courtesy E. I. Zehr)

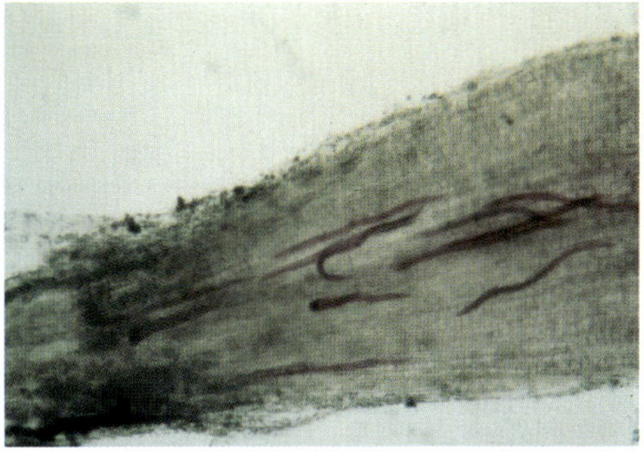

101. *Meloidogyne incognita* in root of susceptible host stained with acid fuchsin. (Courtesy R. S. Hussey)

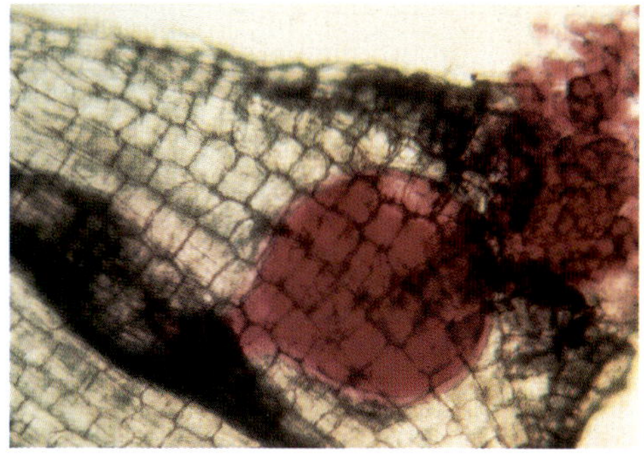

102. Adult female root-knot nematode in root stained with acid fuchsin. Note enlarged root and saccate female producing eggs along the root surface on right. (Courtesy R. S. Hussey)

103. Peach tree infected with root-knot nematode (left) compared with tree on resistant rootstock (right). (Courtesy E. I. Zehr)

104. Necrotic spots and shot-hole symptoms caused by Prunus necrotic ringspot virus. Symptoms appear on cherry leaves during the first year of infection.

105. Delayed fruit maturity (right) caused by rugose mosaic strains of Prunus necrotic ringspot virus.

106. Fruit marking symptoms on Hainier sweet cherry associated with some Prunus necrotic ringspot virus strains.

107. Calico leaf symptoms on cherry caused by some Prunus necrotic ringspot virus strains.

108. Symptoms of sour cherry yellows in Montmorency sour cherry caused by prune dwarf virus. (Courtesy A. L. Jones)

109. Blind wood symptoms (lack of fruiting spurs) on cherry caused by prune dwarf virus.

110. Narrow leaf symptom (right) caused by strains of prune dwarf virus (healthy leaf at left).

111. Symptoms of dense canopy resulting from shortened shoot internodes of peach dwarf disease on peach (cv. Muir).

112. Vein banding symptoms in Kelsey plum caused by a strain of Prunus necrotic ringspot ilarvirus. (Courtesy J. K. Uyemoto)

113. Severe symptoms of cherry mottle leaf on cherry. (Courtesy G. I. Mink)

114. Color break symptoms of peach mosaic disease on peach flowers.

115. Leaf symptoms of peach mosaic disease (chlorotic spots and streaks); healthy leaf at left.

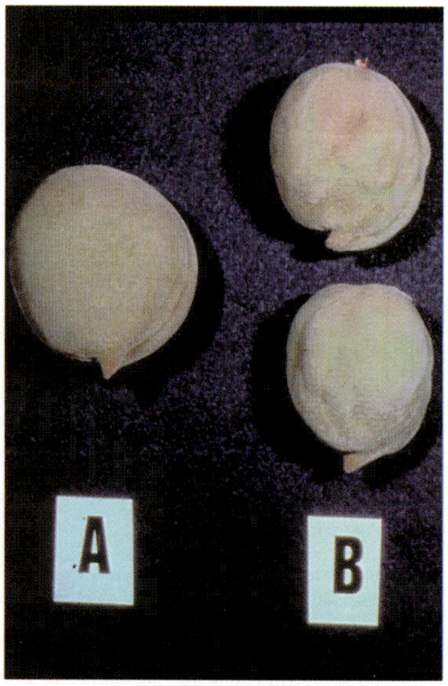

116. Fruit symptoms (deformation and rough surface) of peach mosaic disease on Loring peach (B), compared with a healthy fruit (A).

117. Small, pointed fruit of Bing sweet cherry with little cherry disease (healthy fruit at right). (Courtesy G. I. Mink)

118. Premature fall coloration on cherry (cv. Sam) symptomatic of little cherry disease (healthy cherry tree at right). (Courtesy G. I. Mink)

119. Leaf symptoms (chlorotic spots and veins) caused by plum pox virus on plum (cv. Czar).

120. Fruit symptoms (skin blotches) caused by plum pox virus (plum cv. Czar).

121. Fruit symptoms caused by plum pox virus in apricot. (Courtesy J. A. Foster)

122. Symptoms of chlorotic vein banding and chlorotic oak-leaf patterns in Nanking cherry caused by plum pox virus (healthy leaf at top). (Courtesy V. Damsteegt)

123. Cherry leaf with enation and distortion symptoms caused by cherry rasp leaf virus.

124. Symptoms (shoots with short internodes) caused by peach rosette mosaic virus.

125. Stem pitting on the woody cylinder of a peach rootstock caused by tomato ringspot virus. (Courtesy J. K. Uyemoto)

126. Symptoms of small, yellow leaf clusters (lower vegetative buds) caused by the peach yellow bud mosaic strain of tomato ringspot virus. Normal leaf cluster (far right, upper bud) is larger and green. (Courtesy J. K. Uyemoto)

127. Ringspot symptoms on a peach leaf caused by the peach yellow bud mosaic strain of tomato ringspot virus. (Courtesy J. K. Uyemoto)

128. Symptoms of brown line (partially formed along the scion-rootstock junction) caused by tomato ringspot virus. Host is President plum on peach rootstock. (Courtesy J. K. Uyemoto)

129. Leaf symptoms caused by apple chlorotic leaf spot virus. Note dark green (greasy-appearing) tissues bordering the midrib and lateral veins on infected leaf (healthy leaf at right). Indicator host is seedling peach (cv. Okinawa). (Courtesy M. Barkley)

130. Shot-hole symptoms of apricot ring pox disease on Tilton apricot.

131. Tilton apricot fruit with symptoms of apricot pit pox (right) and apricot ring pox (middle); healthy fruit at left.

132. Cross section through fruit shown in Plate 131. Note necrotic tissues that extend to the pit with apricot pit pox (right) and more superficial necrosis with apricot ring pox (middle).

133. Twisted leaves and chlorotic mottle symptoms caused by the cherry twisted leaf agent.

134. Symptoms of cherry necrotic rusty mottle on Corum cherry. (Courtesy G. I. Mink)

135. Leaf symptoms of cherry rusty mottle disease. (Courtesy J. A. Foster)

136. Short pedicel symptoms of cherry short stem disease in Bing sweet cherry. (Courtesy A. J. Hansen)

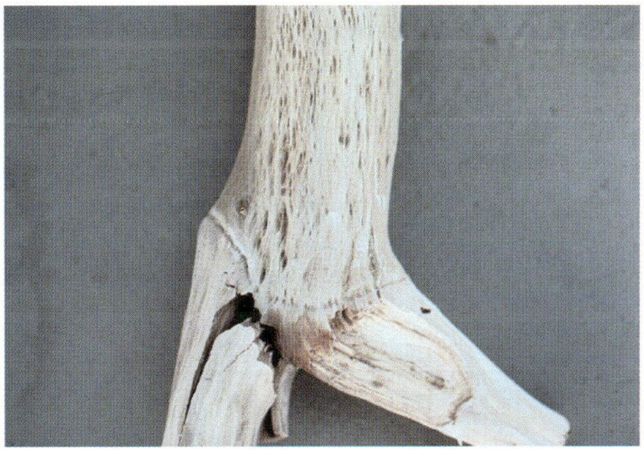

137. Stem pitting symptoms on a three-year-old Bing scion on mahaleb rootstock infected with cherry stem pitting disease.

138. Symptoms of peach mottle disease: yellow mottle and puckering of leaves. (Courtesy A. W. Helton)

139. Crease wart (along suture) on Rio-Oso-Gem peach with peach wart disease. (Courtesy A. L. Jones, from the files of K. G. Parker)

140. Fruit symptoms caused by plum dapple fruit viroid. Healthy, left.

141. Leaf symptoms of (sour cherry) green ring mottle virus in Montmorency sour cherry. (Courtesy A. L. Jones, from the files of K. G. Parker)

142. Leaf chlorosis and midrib necrosis caused by (sour cherry) green ring mottle virus in Shirofugen flowering cherry.

143. Fruit symptoms of sour cherry pink fruit disease on Montmorency sour cherry. Yellowish fruit skins caused by sunburning.

144. Deformed leaves and necrotic fruit symptoms caused by tomato bushy stunt virus on Sam sweet cherry (healthy cherry at right). (Courtesy J. K. Uyemoto)

145. Leaf symptoms of rusty blotch in Santa Rosa plum. Note chlorosis along the leaf margins (healthy leaf at lower right). (Courtesy J. K. Uyemoto)

146. Symptoms of crinkle leaf in Bing sweet cherry. Affected leaves are deformed and irregularly indented. Scattered areas along leaf margins are chlorotic or light green. (Courtesy J. K. Uyemoto)

147. Small fruit symptoms of cherry crinkle leaf. Fruit pedicels vary in length. Healthy fruit at far right. (Courtesy J. K. Uyemoto)

148. Deep suture symptoms shown as indentation along the crease (suture line) on Bing cherry fruit. Two healthy fruit at left. (Courtesy J. K. Uyemoto)

149. Scabbing of prune fruit caused by mechanical leaf rubbing. (Reprinted, by permission, from Ogawa and English, 1991)

150. Symptoms of russet scab on fresh and dried prune fruit. (Reprinted, by permission, from Ogawa and English, 1991)

151. Red protuberances or warts of nectarine pox are generally scattered but occasionally are clustered together.

152. Nectarine pox expressed as small, brown, scabby patches in a random pattern (left) and on the suture (right).

153. Brown and black discoloration on skin of peach fruit. (Courtesy J. D. Ridley)

154. Fasciation of Bing cherry shoots, attributed to genetic mutation of the apical meristem, possibly from natural radiation.

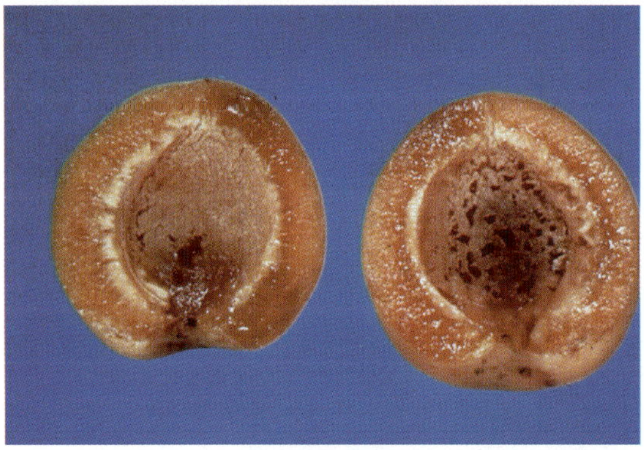

155. Ripe Royal apricots exhibiting a gelatinous inner mesocarp tissue, or pit burning, attributed to excessive respiration rate and lack of oxygen as a consequence of high ambient temperatures.

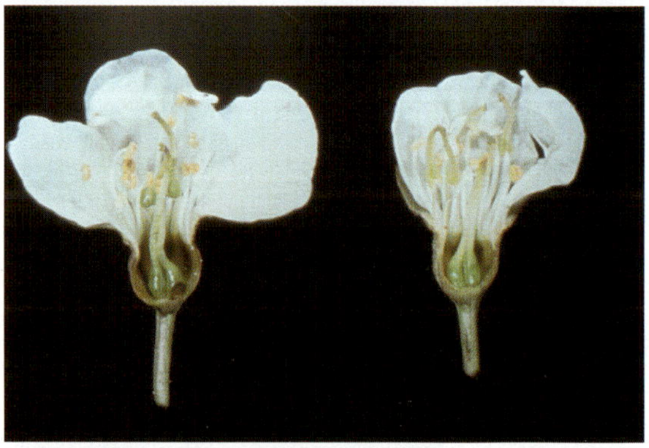

156. Cherry flowers with double pistils. Spurred fruits result when one pistil develops fully while the other develops partially.

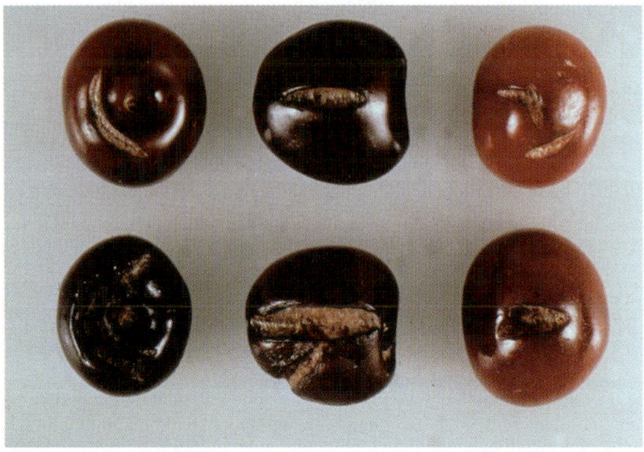

157. Bing cherries exhibiting skin cracks as a consequence of rain during ripening.

158. Side cracking on prunes caused by exposure of fruit to direct sunlight.

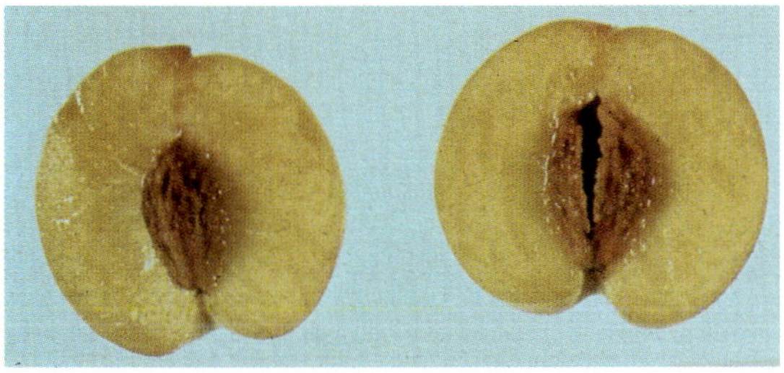

159. Clingstone peach with a split pit (right); normal fruit at left.

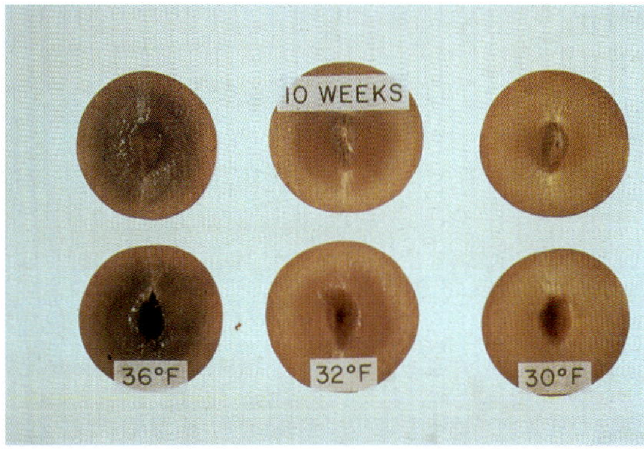

160. Casselman plum with internal breakdown after a long period of cold storage at different temperatures.

161. Nitrogen deficiency on peach. Note pale green leaves, reduced shoot growth, and smaller fruit. Healthy plant at left.

162. Potassium deficiency on prunes. The entire leaf becomes chlorotic, with scorching of margins or the entire blade. These symptoms usually are seen on the tops of the trees and in severe cases lead to dieback of the shoots.

163. Symptoms of magnesium deficiency on prune. Note the inverted green V around the base of the midrib and marginal necrosis.

164. Zinc deficiency (left) on sweet cherries. Note small, chlorotic leaves, shortened internodes, and tufts of small leaves along the shoot (little leaf).

165. Young peach leaves with zinc deficiency. Note chlorosis, interveinal mottling, and crinkled or wavy margins.

166. Iron deficiency on prune leaves expressed as loss of chlorophyll in the interveinal areas, leaving a network of small, green veins.

167. Manganese deficiency on prune leaves. Note general chlorosis with main lateral veins remaining green to produce a herringbone pattern.

168. Copper deficiency of prune. Shoots appear normal in the spring, but shoot terminals become necrotic in early to midsummer.

as pH or drainage adjustments) *or* to invest in specific diagnoses (via bioassays, identification, and population counts) of the pathogens and abiotic causes involved before undertaking appropriate narrow-spectrum, precisely targeted control measures. The costs of preplant corrective measures usually are recovered quickly through earlier and greater fruit production on SFRD sites, but assessment of the site problem should be done first. Assessment of SFRDs is currently done primarily by field trials; use of greenhouse bioassays like those developed for apple replant disease has potential but has not been widely adopted. Most commonly, the problem is ignored until replant trees exhibit slowed growth and delayed fruiting.

Often the fallow period or rotation to another crop needed to control SFRDs takes longer than a grower is willing to wait, and no one alternate crop will work in every situation. Abiotic factors, however, can be alleviated by other cultural methods, such as subsoiling or excavation to improve soil drainage and aeration, correcting soil pH problems and/or fertility problems, and replacing the soil in the planting hole when arsenic toxicity is involved. Phosphorus nutrition is important for new root growth in stone fruits, but application of fertilizer blends (e.g., 11-52-0) has not given responses similar to those obtained for apple replant disease.

Biotic causes typically require soil treatment to change the soil microbiota. Examples include steam pasteurization or sterilization, soil fumigation with general soil sterilants (e.g., chloropicrin, methyl bromide, or metam sodium), or application of nematicides (e.g., 1,3-D, fenamiphos) targeted to the nematode species involved. Soil solarization also has been successful in some instances, but use of fungal antagonists has not proved helpful under normal conditions.

Although subsoiling or excavation to a depth of 1 m can double growth of replant peaches, soil loosening combined with fall preplant fumigation with methyl bromide, chloropicrin, or a combination thereof may improve replant peach growth even more. Preplant soil loosening combined with fumigation also can enable new trees to better withstand drought stress.

Future control choices increasingly will need to be based on prior pest identification and established damage thresholds. Broad-spectrum options such as general biocides face heightened environmental concerns and possible loss of availability. The remaining, more specific control measures need to be integrated into coherent, comprehensive control strategies for each situation.

Selected References

Gur, A., and Cohen, Y. 1988. Causes of soil sickness in replanted peaches: 1. The role of cyanogenesis in peach soil sickness. Acta Hortic. 233:25-31.

Larsen, H. J. 1990. Countering orchard replant problems. Proc. W. Colo. Hortic. Soc. 47:46-56.

Mai, W. F., and Abawi, G. S. 1981. Controlling replant diseases of pome and stone fruits in northeastern United States by preplant fumigation. Plant Dis. 65:859-864.

Traquair, J. A. 1984. Etiology and control of orchard replant problems: A review. Can. J. Plant Pathol. 6:54-62.

Yadava, U. L., and Doud, S. L. 1980. The short life and replant problems of deciduous fruit trees. Pages 1-116 in: J. Janick, ed. Horticultural Reviews, vol. 2. AVI Publishing Co., Westport, CT.

(Prepared by H. J. Larsen)

Parasitic Plants

Mistletoe

Stone fruit trees are parasitized by leafy mistletoes in several genera, including *Phoradendron* and *Viscum* of the family Viscaceae. These evergreen, seed-bearing plants possess chlorophyll but derive additional energy and nutrients from their host. Mistletoes have aerial stems and leaves, rootlike cortical haustoria in host phloem, and structures termed "sinkers" that become progressively embedded in host xylem. This endophytic system removes photosynthates, water, and mineral nutrients from the host vascular system. Leafy mistletoes occur throughout the world on many plants, particularly in tropical and subtropical regions. In the United States, leafy mistletoes have been reported on stone fruit trees from several southeastern states, Texas, Arizona, and California.

Symptoms

Host tissue swells to form a gall at the point of infection by the mistletoe. The stem of the host is extensively invaded, weakened, and may eventually die above the mistletoe infection. Dieback of shoots and reduced tree growth result as more mistletoe plants develop on the tree. Tree death can occur after severe infection but is rarely observed in stone fruit trees. Proliferation of mistletoe shoots forms dense evergreen clusters or clumps in the canopy of the host (Plate 69). These clusters of aerial mistletoe shoots resemble witches'-brooms (proliferations of adventitious shoots of the host) but are actually signs (shoots) of the mistletoe plant.

Causal Organisms

Leafy mistletoe species in the genera *Phoradendron*, *Viscum*, and *Psittacanthus* have been reported on *Prunus* species in North America. *Phoradendron* and *Viscum* spp. occur in the United States, and the following descriptions pertain to these two genera.

Aerial portions of mistletoe plants are evergreen, whereas cortical haustoria and sinkers, which lack chlorophyll, are embedded in the phloem and xylem tissue of the host. The stems of aerial shoots are yellowish green to green and range in length from a few centimeters to more than a meter. The leaves are yellowish green to green, ovate, and opposite. During the dormant season or when infected trees have severe dieback, the mistletoe is conspicuous as distinct clusters of green shoots on the host. Mistletoe plants are dioecious, but male and female plants can occur on the same host. The fruit is a globose, white to pink berry, 3–6 mm in diameter. Each berry contains a single seed that is coated with a sticky, gelatinous pulp.

In North America, several species of leafy mistletoe affect stone fruit trees. In Florida and other southern states, *Phoradendron serotinum* (Raf.) M.C. Johnston (syn. *P. flavescens* (Pursh) Nutt.) has been reported on black (*Prunus serotina* Ehrh.), English (*P. laurocerasus* L.), and American (*P. caroliniana* (Mill.) Aiton) cherries. In western North America, *Phoradendron tomentosum* (DC.) Engelm. ex A. Gray subsp. *macrophyllum* (Engelm.) Wiens occurs on almond (*P. dulcis* (Mill.) D. Webb), plum (*P. americana* Marsh and *P. simonii* Carrière), and prune (*P. domestica* L.), as well as on several species of ornamental and commercially grown cherry trees; *Phoradendron tomentosum* subsp. *tomentosum* (DC.) Engelm. occurs on prune; and *Phoradendron villosum* (Nutt.) Nutt. subsp. *villosum* occurs on peach (*P. persica* (L.) Batsch). In California, *Viscum album* var. *album* L. (European mistletoe) has been reported on almond. In Mexico, *Psittacanthus calyculatus*

(DC.) G. Don of the family Loranthaceae occurs on peach.

The following species of mistletoe have been reported on *Prunus* species in specified regions of the world: *Amyema congener* Tiegh. (Australia); *Dendrophthoe falcata* (L.f.) Ettingsh. (India) and *D. mesembryanthemifolia* Grieb. & Urban (Peru); *Erianthumum dregei* (E. & Z.) Tiegh. (South Africa); *Globimetula braunii* (Engl.) Dans. (Ivory Coast); *Hyphear grewinkii* (Buhse) Dans. (Iran, Pakistan); *Loranthus* sp. (Australia); *Muellerina celastroides* (Seib. ex Schult. & Schult.) Tiegh. (Australia); *Phoradendron quadrangulare* Krug. & Urban (Argentina); *Psittacanthus calyculatus* (DC.) G. Don and *P. cuneifolius* (R. & Pav.) G. Don (Argentina); *Struthanthus oerstedii* (Oliv.) Standl. & Clad. (Costa Rica); *Tapinanthus belvisii* (DC.) Dans. (Ivory Coast); *Taxillus cuneatus* (Roth) Dans. (India); *Tristerix tetrandrus* (Ruiz & Pav.) Mart (Argentina, Chile) and *T. verticillatus* (Ruiz & Pav.) Barlow & Wiens (Argentina); *Viscum album* L. (Europe), *V. articulatum* Burm. (Australia), and *V. farafanganensis* Lecomte (Malagasy Republic).

Disease Cycle and Epidemiology

Leafy or true mistletoes occur mostly on ornamental and forest hardwood tree species and less commonly on stone fruit trees. These parasitic plants are disseminated mainly by birds that eat the seed-containing berries and excrete the seeds onto tree branches. The mucilaginous or gelatinous coating of the seeds allows them to adhere to branches of the host. Seeds can also be carried on the beaks and feet of birds.

A seed germinates after a period of wetness to produce a radicle that grows along the branch. A haustorium penetrates the bark and grows into the living phloem and cambium; sinkers are formed from the haustorium. As the host grows, the sinkers elongate in the cambial region and become embedded in the wood while maintaining their connection to the haustorium in the cambium. Haustoria and sinkers function as anchors for the mistletoe plant and for absorbing water, minerals, and some carbohydrates from the host. Aerial shoots and leaves are produced at the original point of infection and expand outward concentrically as the plant grows. Flowering and production of fruit occur on mature shoots, which are perennial and function for several years.

Mature, large trees are commonly infected because birds prefer to perch on the tops and in the crown of tall trees. Infections often become severe when birds spend prolonged periods in these trees feeding on mistletoe fruit and excreting the seeds onto branches of the tree. The mistletoe plant removes nutrients and water from the host, weakening it, and eventually may kill the branch beyond the point of infection. Numerous infections may weaken the entire tree, predispose it to other diseases, insects, and disorders, or cause death.

Control

Pruning infected branches that contain aerial and internal portions of the parasitic plant, as well as removing heavily infected trees, can control the disease and prevent new infections of healthy branches and trees in the orchard. Direct removal of mistletoe shoots reduces stress on the host and prevents localized spread of mistletoe by reducing the seed source. Removal of only the aerial portion of the mistletoe does not prevent development of new aerial shoots from haustoria embedded in the host branch. After removal of mistletoe shoots, some methods prescribe covering the affected host branch with creosote or opaque materials, such as tar paper or black plastic sheeting. These practices have not been particularly effective.

Pruning infected branches or trees may not be practical when infections occur on major scaffold branches, trunks, or ornamental fruit trees. Some success has been obtained with chemical treatments sprayed on other hardwood species. The isopropyl ester of 2,4-D, amitrol, or atrazine mixed with dormant oils provided some control of mistletoe on walnuts for five years after treatment. Other mixtures of 2,4-D, dicamba, and cycloheximide significantly reduced regrowth of mistletoe shoots on several hardwood ornamental species for 1.5 years after treatment. Application of ethephon causes early abscission of aerial shoots but does not kill the endophytic system, and more shoots eventually develop. Although eradication is difficult and seldom obtained with chemical treatments, delayed regrowth of shoots reduces seed production and limits secondary spread of these parasitic plants.

Selected References

Bayer, D. E., Elmore, C. L., Chaney, D. H., and Buschmann, L. L. 1967. Control of mistletoe, *Phoradendron flavescens*, on English walnuts in California. HortScience 2:10-11.

Hawksworth, F. G. 1974. Mistletoes on introduced trees of the world. U.S. Dep. Agric. For. Serv. Handb. 469. 49 pp.

Scharpf, R. F., and Hawksworth, F. G. 1974. Mistletoes on hardwoods in the United States. U.S. Dep. Agric. For. Serv. For. Pest Leafl. 147. 7 pp.

Wiens, D. 1964. Revision of acatophyllous species of *Phoradendron*. Brittonia 16: 11-54.

(Prepared by J. E. Adaskaveg and R. L. Gilbertson)

Diseases Caused by Bacteria

Bacterial Canker

Bacterial canker of stone fruit occurs in all major fruit-producing areas of the world. The disease is also known as gummosis, blossom blast, dieback, spur blight, and twig blight. Studies on bacterial canker and development of disease control strategies are complicated by the wide host range of the pathogens, systemic invasion of host tissue without visible symptoms, and epiphytic growth of pathogens on host and nonhost leaf surfaces of plants.

The severity of damage varies from subtle, almost undetectable effects to rapid death of trees in nurseries and orchards. The disease can have a significant effect on fruit production on mature trees. A combination of several predisposing elements, particularly climatic and soil factors, favors disease expression.

Symptoms

Disease symptoms depend on cultivar, age of tree, plant tissue invaded, strain of the pathogen, and nature of the predisposing factors. Cankers, the most characteristic symptom, may not appear on some hosts. Cankers develop on twigs at the base of flower and leaf buds, in pruning wounds (including those made during removal of suckers), and at the base of infected spurs. Cankers generally spread upward, and the infected, sunken areas are usually formed in late winter or early spring. Gum often exudes from cankers, especially early in the growing season. Terminal shoots or twigs of cankered trees may die back. If girdled by a canker, the diseased lateral branch or trunk dies within weeks. However, the root system of diseased trees usually remains healthy, and suckers develop in the crown region.

The pathogens may be present in dormant leaf and flower

buds. Infected, dormant buds often are killed (Plate 70), but some invaded buds open normally in spring, only to collapse in early summer. Leaves from these buds wilt, and fruit tends to dry out. In contrast, leaves and flowers arising from other infected buds may remain symptomless.

Leaf infections, especially on cherry, appear as water-soaked spots that later become brown and dry (Plate 71). At a later stage, shot holes may be seen (Plate 72, left side). However, symptoms on leaves occur sporadically and are therefore not always typical of the disease. If blossom infection occurs, cankers subsequently form on twigs and spurs, and the dead flowers remain attached on fruit spurs (Plate 73). Flat, superficial, dark brown spots develop on infected fruit (Plate 72, right side); however, lesions may be depressed into the fruit flesh, especially when immature cherry fruit are infected.

Causal Organisms

Two pathovars of *Pseudomonas syringae* may cause bacterial canker of stone fruits: *P. syringae* pv. *syringae* van Hall may cause cankers on any fruit species grown commercially, whereas *P. syringae* pv. *morsprunorum* (Wormald) Young et al infects predominantly cherry (sweet and sour) and plum. The behavior of these two pathogens in host tissue is very similar. A third pathovar, *P. syringae* pv. *persicae* (Prunier et al) Young et al, causes leaf spots, cankers, and gummosis of fruit on peach in France and bacterial decline of nectarine, peach, and Japanese plum (*Prunus salicina*) in New Zealand.

P. syringae pv. *syringae* and *P. syringae* pv. *morsprunorum* are placed in RNA group I of the genus *Pseudomonas*. They are gram-negative, straight rods measuring 0.7–1.2 × 1.5 µm and are motile by more than one polar flagellum. They do not accumulate poly-β-hydroxybutyrate as a carbon reserve and are negative for arginine dihydrolase and oxidase. Most strains produce fluorescent pigments on King's medium B. After 72 hr of incubation at 26°C, colonies on this medium are circular with entire or lobate margins, umbonate or umbilicate, smooth, and glistening and appear vitreous against the light. Levan-forming strains produce large, whitish to cream-colored, dome-shaped colonies on nutrient sucrose agar.

Several procedures have been used to distinguish the two pathovars. Of these, the GATTa scheme (for gelatin liquefaction, aesculin hydrolysis, tyrosinase activity, and tartrate utilization) is recommended. *P. syringae* pv. *syringae* and *P. syringae* pv. *morsprunorum* score $G^+A^+T^-Ta^-$ and $G^-A^-T^+Ta^+$, respectively, but some intermediate strains diverge for one or more of the four tests. Most strains of *P. syringae* pv. *syringae* produce the toxin syringomycin. An ice-nucleation test can further differentiate pathovars; *P. syringae* pv. *morsprunorum* has no or very little ice-nucleation activity, whereas most strains of *P. syringae* pv. *syringae* act as ice-nucleating agents. Phenotypic heterogeneity of *P. syringae* pv. *syringae,* and to a lesser extent *P. syringae* pv. *morsprunorum,* appears to be more extensive than some diagnostic schemes indicate. Nevertheless, the existence of two separate pathovars has been justified by DNA-DNA hybridization studies. *P. syringae* pv. *persicae* is closely related phenotypically to *P. syringae* pv. *morsprunorum.*

The hypersensitive reaction on tobacco leaves is used as a presumptive test for pathogenicity. Effective bioassays based on the response of inoculated immature cherry fruit or peach seedling cotyledons have been developed for detecting pathogenicity of *P. syringae* pv. *syringae* strains. This method has the important advantage of distinguishing pathogenic strains obtained from a wide range of hosts. Virulence can be measured by inoculating etiolated pear and apple hypocotyls.

Disease Cycle and Epidemiology

P. syringae pv. *syringae* and *P. syringae* pv. *morsprunorum* overwinter in cankers and buds and systemically inside other symptomless host tissue. The pathogens colonize new leaves in spring as they emerge from invaded buds. *P. syringae* pv. *syringae* also grows epiphytically on weeds and leaves of crop plants. Bacterial populations on leaf surfaces provide an important source of inoculum for infections of susceptible hosts. In late winter, periods of frequent rainfall, high humidity, cool temperatures, and wind favor dispersal and infection by the pathogens. Substomatal chambers on host leaves subsequently serve as protected sites that enable the pathogens to survive adverse conditions during warmer, drier periods. Rain during the growing season ensures that inoculum is dispersed throughout the orchard.

After entering through stomata, the pathogens colonize intercellular spaces of the spongy parenchyma. When substomatal chambers are reached, bacteria can multiply profusely, and masses of bacterial cells are extruded through stomata. Epiphytic populations of the pathogen may be replenished constantly in this way. The pathogen also moves through the parenchyma of the bundle sheath to axillary buds and to twigs supporting the leaves.

The trunk and branches of trees usually are infected in autumn and winter. Pruning wounds and sites of freeze injury provide places of entry. In England, most cankers caused by *P. syringae* pv. *morsprunorum* on sweet cherry trees originate during autumn, but both pathovars also may reach axillary buds by systemic spread from the leaves through the petiole well before leaf fall. Cankers subsequently appear at the base of invaded buds. Canker expansion slows during the winter but resumes again in the spring. Further development may be arrested later in the spring when callus tissue forms. Populations of the pathogens in cankers decline during summer. Cankers can become inactive after one year or, if the pathogen is not contained effectively, resume expansion.

Blossom infection occurs in some orchards. Lesions on fruit are caused by secondary spread of bacteria oozing from leaf spots and cankers. Under experimental conditions, *P. syringae* pv. *morsprunorum* on sweet cherry blossoms has spread systemically through the developing fruit into twigs to initiate cankers. However, not all infected flowers are killed, and recent evidence indicates that *P. syringae* pv. *syringae* can be transmitted to peach seed used for rootstocks. Some symptomless rootstocks thus may carry latent populations of the pathogen originally present in seed.

Control

Trees with latent infections probably occur in most major fruit-producing regions of the world. However, if predisposing factors are avoided or counteracted, bacterial canker is unlikely to be a serious problem. Some of the stress factors that have been recognized in the United States and elsewhere include freeze injury, wounds, nematode damage, and coincident infection by *P. syringae* pv. *syringae* and plant-pathogenic fungi such as *Leucostoma* (*Cytospora*) spp. and *Nectria* spp.

New orchards established under marginal climatic and soil conditions are at risk. Trees are particularly susceptible in sandy soils, in waterlogged soils that drain poorly, and during prolonged periods of drought. Also, cultivar and species should be selected to ensure that the dormancy requirement is properly met in regions that experience mild winters. Precautions should be taken during pruning to limit spread of the pathogens. Pruning in early summer, rather than autumn and winter, lessens the chance that trees will become infected. It has been claimed that periodic removal of cankers on cherry trees helps reduce disease incidence.

Major outbreaks of bacterial canker in young orchards often are attributed to poor cultural practices. The situation is undoubtedly aggravated if the pathogen has been introduced uniformly into young nursery trees through infected buds or rootstocks. These trees may not have become diseased in nurseries where optimal growing conditions were maintained. However, the stress of transplanting and less-than-optimal conditions in the orchard site may make the trees more prone to infection until they become established. Selection of proper rootstocks and cultivars for a particular geographic region is important. In California, for example, plums on Lovell peach rootstock and French prune on both Lovell and Nemaguard rootstocks sustain less damage from bacterial canker than do those on Myrobalan or Marianna rootstocks.

Chemical control of bacterial canker in Europe and North America is based primarily on protective copper sprays in autumn and in spring before flowering. These sprays protect trees from initial infection but cannot prevent the canker phase once infection has occurred. Copper resistance occurs in *P. syringae* pv. *syringae* isolated from orchards in Michigan, California, and Oklahoma. Copper-resistant strains can survive from one season to the next in the presence of copper applications, which potentially reduces the effectiveness of these treatments.

Selected References

Cameron, H. R. 1962. Disease of deciduous fruit trees incited by *Pseudomonas syringae* van Hall. Oreg. Agric. Exp. Stn. Tech. Bull. 66. 64 pp.

Crosse, J. E. 1966. Epidemiological relations of the pseudomonad pathogens of deciduous fruit trees. Annu. Rev. Phytopathol. 4:291-310.

Endert, E., and Ritchie, D. F. 1984. Detection of pathogenicity, measurement of virulence, and determination of strain variation in *Pseudomonas syringae* pv. *syringae*. Plant Dis. 68:677-680.

Hattingh, M. J., Roos, I. M. M., and Mansvelt, E. L. 1989. Infection and systemic invasion of deciduous fruit trees by *Pseudomonas syringae* in South Africa. Plant Dis. 73:784-789.

Latorre, B. A., and Jones, A. L. 1979. *Pseudomonas morsprunorum*, the cause of bacterial canker of sour cherry in Michigan, and its epiphytic association with *P. syringae*. Phytopathology 69:335-339.

Ogawa, J. M., and English, H. 1991. Diseases of Temperate Zone Tree Fruit and Nut Crops. Publ. 3345. University of California Division of Agriculture and Natural Resources, Oakland. 461 pp.

Sundin, G. W., Jones, A. L., and Fulbright, D. W. 1989. Copper resistance in *Pseudomonas syringae* pv. *syringae* from cherry orchards and its associated transfer in vitro and in planta with a plasmid. Phytopathology 79:861-865.

(Prepared by M. J. Hattingh and I. M. M. Roos)

Bacterial Decline

Bacterial decline has been recorded on nectarine and peach in France and on nectarine, peach, and Japanese plum (*Prunus salicina*) in New Zealand. The same pathogen may cause a bacterial decline of Myrobalan plum (*P. cerasifera*) in the United Kingdom. This disease can be considered as part of the bacterial canker complex discussed in the preceding section.

Symptoms

In nectarine and peach, symptoms include shoot dieback, limb and root injury, tree death, leaf spots, and fruit lesions. On Japanese plum, symptoms are mainly confined to dieback, occasional limb death, and leaf spots. Dieback of shoots occurs in autumn, on the current season's growth, and in spring, following the development of girdling lesions from nodal infections. Small, elliptical lesions may develop at internodes. In spring and early summer, wilting of flowers and leaves on branches or on entire trees indicates the development of extensive necrosis in the main leaders or trunk. Infection is confined to cambium and phloem tissue of the bark (Plate 74). The rootstock and roots also can be infected. Leaf infection results in small, angular, water-soaked spots, the tissue of which becomes brown and falls out to produce shot holes. Small, round, dark, oily spots occur on fruit. These can spread within the fruit tissue, causing sunken, deforming lesions that ooze gum (Plate 75).

Some symptoms of bacterial decline can be confused with those of bacterial canker. Distinctive characteristics of decline are the staining of wood in branches above the necrosis and the absence of an obvious boundary between the morbid and healthy bark in the lower parts of the tree.

Cultivars appear to differ in susceptibility; some exhibit partial resistance, although none are immune.

Causal Organism

Pseudomonas syringae pv. *persicae* (Prunier et al) Young et al can be distinguished by several determinative tests from other pathovars of *P. syringae* that attack stone fruit. The pathogen has the following LOPAT characteristics: it produces levan colonies, is negative for oxidase, pectolytic activity, and arginine dihydrolase, and elicits tobacco hypersitivity. Some strains produce small amounts of blue fluorescent pigment on a suitable medium such as King's medium B. On this medium, the initial colony growth is gray and translucent; it increases to 2–3 mm in diameter and becomes creamy and opaque.

Disease Cycle and Epidemiology

The pathogen can become established in trees in the nursery; thereafter, it is continually present in trees, surviving in summer in or on symptomless leaves.

The decline syndrome has affinities with the conditions described as "apoplexy of apricot" and peach tree short life, which also involve bacteria. Disease development is mainly associated with cold weather. In autumn, dieback is severe if early frosts occur before trees have hardened off. Pathogenic activity is greatest in winter and early spring following infection through leaf scars and perhaps through frost injury or pruning cuts. Extensive invasion of bark, often extending more than 1 m along branches, occurs at this time. Bacterial infection and tree damage and death can be most severe when a soil condition, such as waterlogging, causes root injury. Root infection appears to occur in late winter or spring and is not immediately associated with the development of obvious necrosis. During summer, the pathogen can be isolated from necrotic bark.

Control

No completely satisfactory control program has been developed. Preventive measures include avoidance of cultural practices that lead to root injury. Other measures that reduce disease occurrence are the use of disease-free nursery stock, selection of cultivars with partial resistance, avoidance of early-winter pruning, and application of fixed-copper sprays in autumn and winter to reduce inoculum.

Selected References

Luisetti, J., Prunier, J. P., Gardan, L., Gaignard, J. L., and Vigouroux, A. 1976. Le dépérissement bactérien du pecher (*Pseudomonas mors-prunorum* f. sp. *persicae*). Institut national de vulgarization pour les fruits, legumes et champignons (Invulflec), Paris. 60 pp.

Young, J. M. 1988. *Pseudomonas syringae* pv. *persicae* from nectarine, peach, and Japanese plum in New Zealand. Bull. OEPP 18:141-151.

(Prepared by J. M. Young)

Bacterial Spot

Bacterial spot was originally described by Erwin F. Smith in 1903 on Japanese plum (*Prunus salicina* Lindl.) from Michigan. The disease also is referred to as bacteriosis, bacterial leaf spot, bacterial shot hole, bacterial crack, and black spot. Bacterial spot occurs in most countries where stone fruits are grown, including, but not limited to, Australia, Canada, Japan, Italy, New Zealand, South Africa, the former Soviet Union, the United States, Brazil, and several other South American countries. In the United States, bacterial spot occurs in all stone fruit-growing areas east of the Rocky Mountains; it has not been reported in the Pacific Coast area. Bacterial spot is more common and most severe in areas where stone fruits are grown in light, sandy soils and the environment is humid or moist and warm during the growing season. The most common hosts include peach and nectarine, Japanese plum, apricot, and almond. Other hosts are sweet and tart cherries, Japanese apricot (*P. mume* Sieb. & Succ.), Chinese wild peach (*P. davidiana* (Carrière) Franch.), *P. buergeriana, P. crassipes,* and *P. donarium.* American and European plum are less susceptible than Japanese plum. Cultivars within all of these *Prunus* species vary widely in susceptibility.

Symptoms

Bacterial spot occurs on leaves, twigs, and fruit. Leaf symptoms are first visible as angular, grayish, water-soaked lesions, approximately 1–3 mm in diameter, often located along the midrib, leaf tip, or both, or along the leaf margin. As the lesions age and enlarge to several millimeters in diameter, lesion centers become purple and necrotic, and if they abscise, a shot-hole appearance results. Multiple lesions result in leaf chlorosis and premature abscission (Plate 76). Severe defoliation can occur on highly susceptible cultivars. Consecutive years of disease with severe premature defoliation may weaken trees and reduce fruit quality. Initial leaf and fruit symptoms may be confused with pesticide injury.

Several types of lesions occur on twigs and are designated as "spring" or "summer" cankers or as "black tip." Spring cankers occur on twigs of the previous summer's growth, developing from leaf scar infections in late autumn. Visible about the time of leaf emergence, spring cankers appear as slightly raised, blisterlike areas that can extend several centimeters along the twig (Plate 77). Summer cankers are formed on new green shoots and are visible first in late spring or early summer. Black tip is visible in late winter before leaf emergence and is limited to the terminal bud region of the previous year's growth. The terminal bud usually fails to open, and a dark cankered area, completely surrounding the twig, can extend several centimeters downward from the tip (Plate 78). Both spring cankers and black tip result from infections the previous autumn. Dieback as a result of twig cankers is more severe on plum and apricot than on peach.

On highly susceptible cultivars, the entire fruit crop can be lost in years when bacterial spot is severe. Fruit symptoms first are visible about three to five weeks after petal fall as small, water-soaked, brownish lesions (Plate 79). Early symptoms on fruit can be mistaken for insect damage. During periods of high humidity, gum may exude from these lesions (Plate 80). As the season progresses, lesions become cracked and sunken and may even appear "cavernous." The frequency and severity of disease on fruit does not always correlate with that on leaves.

Causal Organism

Erwin F. Smith originally named the bacterial spot pathogen *Pseudomonas pruni* E.F. Smith in 1903. Subsequent names were *Bacterium pruni* E.F. Smith, *Phytomonas pruni* (E.F. Smith) Bergey et al, and *Xanthomonas pruni* (E.F. Smith) Dowson. The currently accepted name is *Xanthomonas campestris* pv. *pruni* (Smith) Dye. The bacterium is a gram-negative, motile rod measuring 0.2–0.4 × 0.8–1.0 µm. It is a strict aerobe with an optimum growth temperature range of 24–29°C. Growth on yeast-dextrose-calcium carbonate medium at 28°C for 48–72 hr results in mucoid, yellow-pigmented colonies. Selective and differential media such as that of Gitaitis et al have been developed for isolation of this pathogen from plant material. Differential virulence among strains of *X. campestris* pv. *pruni* to peach, plum, and apricot cultivars occurs.

Disease Cycle and Epidemiology

In autumn, the pathogen invades peach twigs via fresh leaf scars. These infections then are expressed as spring cankers or black tip in the following winter to early spring. In plum, the major route of twig infections appears to be by systemic movement of bacteria through leaf petioles from infected leaves. In peach, most primary inoculum is believed to be overwintered

TABLE 3. Relative Susceptibility of Selected Peach, Nectarine, and Japanese-type Plum Cultivars to Bacterial Spot

Fruit	Resistant	Susceptible	Highly Susceptible
Peach[a]			
	Biscoe	Bellaire	Autumn Lady
	Bounty	Bicentennial	Calred
	Candor	Carolina Belle	Carnival
	Clayton	Contender	Cullinan
	Derby	Cresthaven	Elberta
	Dixired	Encore	Elegant Lady
	Jerseydawn	Ernie's Choice	Fairtime
	Newhaven	Gabriella	Fireprince
	Salem	Garnet Beauty	Firered
	Sentinel	Gaucho	Marsun
	Sweethaven	Grazano	May Lady
		Harrow Beauty	Monroe
		Harrow Diamond	O'Henry
		Harvester	Parade
		Jayhaven	Royal May
		Jerseyqueen	Ryansun
		Legend	Scarletpearl
		Loring	Summerglo
		Majestic	Summerpearl
		Maygold	Sunbrite
		Norman	Suncrest
		Ouachita Gold	Sunhigh
		Raritan	Surecrop
		Red Haven	Sweet Sue
		Redskin	
		Rio Oso Gem	
		RoseRedglobe	
		Ruston Red	
		Summergold	
		Sunprince	
		Topaz	
		Winblo	
Nectarine[a]			
	Carolina Red	Armking	Cassiopia
	Le Grand	Earlibird	Crimson Gold
	Nectared 3	Fantasia	Firebrite
	Nectared 6	Hardired	Flamekist
	Royalkist	Roseprince	Flavortop
	Sweet Melody	Royal Giant	Lagold
		Snow Queen	Late Gold
		Summer Beaut	Redgold
		Sunfre	
		Sunglo	
		Sungrand	
		Tasty Free	
Japanese plum[b]			
	AU Amber[c]	AU Producer[c]	AU Frontier M1[c]
	AU Crimson[c]	Beauty	AU Rosa[c]
	Bellerosa	Blackamber	AU Rubrum[c]
	Bruce	Burbank	Frontier[c]
	Earlisweet	Burmosa	Methley[c]
	Explorer	Byrongold[c]	Santa Rosa[c]
	Homeside	Eldorado	
	Queensland	Friar	
	Robusta	Laroda	
		Morris[c]	
		Nubiana	
		Ozark Premier[c]	
		Queen Rosa	
		Redheart	
		Rubysweet[c]	

[a] Some of the information was adapted from Werner et al (1986) and unpublished observations in North Carolina.
[b] Some of the information was adapted from Topp and Sherman (1990).
[c] Data from unpublished observations in 1993 in North Carolina.

bacteria associated with spring cankers and black tip. However, bacteria may overwinter in terminal buds of peach and also have been detected in epiphytic association on twigs and buds. If epiphytic survival is proved to be prevalent, cankers may not be essential as overwintering sites. The pathogen has been readily isolated from lateral buds in summer and autumn. On plum and apricot, bacteria can overwinter in summer cankers, and the cankers continue to develop the following spring. The occurrence and development of primary and subsequent secondary infections depend entirely on environmental conditions. Frequent periods of moisture during late bloom to a few weeks after petal fall are very conducive to primary fruit and leaf infections on peach and nectarine. Wind-driven rain may increase disease severity. Similar environmental conditions throughout the growing season allow for the continuation of secondary infections. Few infections occur during hot, dry conditions.

Control

Where bacterial spot occurs, planting highly susceptible cultivars should be avoided. Bacterial spot is severe on apricot cultivars Blenril, Early Golden, and Veecot but mild to moderate on Newcastle, Stella, Superb, Viceroy, and Moorpark. When environmental conditions are highly conducive to disease development, no cultivar is immune, and the severity of disease can vary yearly on the same cultivar. In recent years, however, cultivars have been developed that have some degree of bacterial spot resistance (Table 3).

Efforts should be made to avoid introduction of bacterial spot into new stone fruit-growing areas where environmental conditions are favorable for disease development. Because the pathogen can survive in lateral buds during the summer and early autumn, use of budwood from orchards with diseased trees should be avoided. If susceptible cultivars are planted, they should not be planted in adjacent blocks. Adequate fertility should be maintained to avoid excessive foliar growth or weakening trees from poor nutrition. Both of these factors enhance bacterial spot development.

Successful chemical control may depend highly on timing of applications. Once the disease is observed, it is difficult to control. Autumn applications near leaf drop and/or early-season applications of fixed copper may be beneficial in preventing leaf scar infections and reducing overwintering inoculum on the tree surface and in inhibiting bacterial movement from overwintering cankers to newly emerging leaves and fruit. Chemical sprays applied at bloom and for several weeks after bloom have reduced the amount of diseased fruit at harvest. Chemicals other than copper that have been used with varying success are oxytetracycline, dodine combined with captan, and zinc-containing chemicals such as ziram and zinc sulfate. Because peaches and some other stone fruits are very sensitive to copper, its use normally is limited to dormant and early-season sprays, although limited success has been obtained with very low rates of copper used during the growing season.

Selected References

du Plessis, H. J. 1988. Differential virulence of *Xanthomonas campestris* pv. *pruni* to peach, plum, and apricot cultivars. Phytopathology 78:1312-1315.

Gitaitis, R. D., Hamm, J. D., and Bertrand, P. F. 1988. Differentiation of *Xanthomonas campestris* pv. *pruni* from other yellow-pigmented bacteria by the refractive quality of bacterial colonies on an agar medium. Plant Dis. 72:416-417.

Ritchie, D. F., and Bennett, M. H. 1993. Use of early season sprays to manage bacterial spot on fruit of peaches, 1992. Fungic. Nematic. Tests 48:62.

Shepard, D. P., and Zehr, E. I. 1986. Epiphytic behavior of *Xanthomonas campestris* pv. *pruni* on peach and plum. (Abstr.) Phytopathology 76:1070.

Thornberry, H. H., and Anderson, H. W. 1933. Overwintering of *Phytomonas pruni* on peach. Phytopathology 23:787-801.

Topp, B. L., and Sherman, W. B. 1990. Sources of bacterial spot resistance in Japanese-type plum cultivars. Fruit Var. J. 44(1):32-35.

Werner, D. J., Ritchie, D. F., Cain, D. W., and Zehr, E. I. 1986. Susceptibility of peaches and nectarines, plant introductions, and other *Prunus* species to bacterial spot. HortScience 21:127-130.

(Prepared by D. F. Ritchie)

Crown Gall

Crown gall is a widespread and important bacterial disease that affects more than 200 dicotyledonous herbaceous and woody plant species. *Prunus* spp. may be heavily infected in the nursery or as mature trees in the orchard. Diseased plants in the nursery are culled or sold at a reduced price, resulting in significant economic losses for the nursery owner. In a related disease named hairy root, infected plants produce excessive amounts of fibrous roots. Hairy root is primarily of concern on apple and develops only infrequently on *Prunus*.

Symptoms

Galls develop on roots, crowns, and occasionally aerial parts of stone fruit trees (Plate 81). They vary in size from microscopic to more than 10 cm in diameter. The galls are composed of disorganized vascular and parenchyma tissues and appear first as rapidly growing, white, fleshy swellings that usually become tan to brownish during the dormant season. New galls frequently develop adjacent to those from the previous year. Galls develop at wound sites and, when small, can be confused with normal callus tissue. Latent infections remain unnoticed during storage and grading of nursery plants, but after planting in the field, these infections can become active and result in gall formation.

Reports vary as to the effects of crown gall on plant growth. Some indicate that galls on young plants can reduce root development and plant vigor, but in at least one experiment on cherry, no significant adverse effect was detected. Heavily galled plants in nurseries may be stunted and show chlorotic leaf growth. However, crown gall may remain undetected for years in mature orchards and become apparent only when removal of trees exposes the diseased roots.

Causal Organism

The crown gall pathogen, *Agrobacterium tumefaciens* (Smith & Townsend) Conn, and the hairy root pathogen, *A. rhizogenes* (Riker et al) Conn, belong to the family Rhizobiaceae and are gram-negative, aerobic, motile, rod-shaped (0.6–1.0 × 1.5–3.0 μm) bacteria that lack endospores. *A. tumefaciens* is divided into two groups or biovars. Biovar 2 is commonly found associated with the disease on stone fruits. Selective culture media have been developed for isolation of the bacterium from plants and soil. Nonpathogenic strains are classified as a different species, *A. radiobacter* (Beijerinck & van Delden) Conn.

The ability of *A. tumefaciens* to cause galls is associated with genes that are present on a tumor-inducing (Ti) plasmid. When the bacterium infects, a segment of the Ti plasmid (the T-DNA) is transferred to the plant and becomes incorporated in the plant chromosome. In response, the host cell overproduces plant hormones, thereby initiating rapid and uncontrolled cell growth that results in gall formation. The T-DNA also directs the cells to produce unusual amino acids (opines) which are utilized by the pathogen. Opines impart a nutritional advantage to the pathogen over other microorganisms in the gall tissue. The Ti plasmid may be transferred from *A. tumefaciens* to *A. radiobacter* in gall tissue, thus changing the nonpathogenic species into a pathogen. A similar genetic mechanism of disease exists for hairy root, but in this case the genetic determinants are located on the Ri plasmid.

Disease Cycle and Epidemiology

Agrobacterium spp. are commonly isolated from field soil. Almost all strains recovered are nonpathogenic (*A. radiobacter*), but pathogenic strains may be isolated from soil, from galls, or from apparently healthy roots of host plants. The bacteria may be present in latent infections, on the surface of roots, or in soil adhering to root surfaces. Therefore, the pathogen may be disseminated with plants or by movement of soil or water. *A. tumefaciens* may persist in field soil for at least one year, or considerably longer if large amounts of infected root residues remain in soil when trees are removed.

Crown gall infections are initiated at wounds. The most susceptible wounds on stone fruit trees are typically at the crown or on the roots and rarely on trunks and limbs. The bacteria first attach to wounded host cells and then transfer T-DNA into the cell. Most stone fruits are very susceptible to crown gall. Infections are favored by moist, alkaline, poorly drained soils and can be stimulated by the feeding damage of plant-parasitic nematodes. Previous cropping history as it relates to inoculum levels in soil affects disease development. When temperatures are above 20°C, galls first may be obvious two to four weeks after infection. At lower temperatures, infections and disease development are delayed, and latent infections may not become apparent for the duration of the dormant season.

Control

Chemical control for crown gall has been either ineffective or impractical for use on a commercial scale. Attempts to control crown gall by soil fumigation also have failed. A biological control agent developed in Australia provides excellent control of the disease on stone fruits. It utilizes a nonpathogenic strain, *A. radiobacter* strain K84, that produces the antibiotic agrocin 84. A recently developed derivative of strain K84, designated K1026, is superior to the original K84 strain. Agrocin 84 inhibits or kills many strains of *A. tumefaciens* (such as those that infect stone fruits), and disease is prevented. K84 also competes for infection sites and therefore can block infection by the pathogen. Roots and crowns of trees are dipped into a suspension of the biological control bacterium before they are planted into field soil. The control procedure has worked successfully on stone fruits in many parts of the world. Commercial preparations are available in several countries under different trade names. Once galls are formed, bacticin (a hydrocarbon containing 2,4-xylenol and metacresol) provides a therapeutic treatment for the galls.

Sanitation and cultural practices also are important for reducing the amount of disease. These include using certified disease-free propagative material, planting disease-free plants, planting in well-drained fields, and rotating contaminated field sites with nonhost plants such as monocots.

Selected References

Garrett, C. M. E. 1987. The effect of crown gall on growth of cherry trees. Plant Pathol. 36:339-345.

Jones, D. A., and Kerr, A. 1989. *Agrobacterium radiobacter* strain K1026, a genetically engineered derivative of strain K84, for biological control of crown gall. Plant Dis. 73:15-18.

Kerr, A. 1972. Biological control of crown gall: Seed inoculation. J. Appl. Bacteriol. 35:493-497.

Moore, L. W. 1976. Latent infections and seasonal variability of crown gall development in seedlings of three *Prunus* species. Phytopathology 66:1097-1101.

Schroth, M. N., and Hildebrand, D. C. 1968. A chemotherapeutic treatment for selectively eradicating crown gall and olive knot neoplasms. Phytopathology 58:848-854.

Schroth, M. N., and Moller, W. J. 1976. Crown gall controlled in the field with a nonpathogenic bacterium. Plant Dis. Rep. 60:275-278.

(Prepared by T. J. Burr)

Phony Peach

Phony (phoney) peach disease (PPD) has been endemic in the southeastern United States since at least 1890, when it was first observed in an orchard in Marshallville, Georgia, and considered a curiosity. Epidemics erupted in 1915–1929 and 1971–1978 in orchards from eastern Texas through South Carolina, including northern Florida and southern Arkansas, Tennessee, and North Carolina. The disease was particularly severe in Florida, southern Alabama, and Georgia. Between and since these epidemics, the disease has been quiescent or has occurred in a restricted range at infection levels that were easily managed.

Symptoms

The canopy of peach trees infected with the PPD pathogen is flattened, compact, and umbrella-like as a consequence of shortened stem internodes, and the foliage is unusually dense and dark green (Plate 82). Dwarfing is conspicuous in symptomatic young trees. Other symptoms include early bloom and fruit set and reduced fruit size. A tree infected for several years produces fruit that is generally half normal size, and production may be reduced 80–90% because the bearing wood fails to develop.

Causal Organism

PPD is caused by a gram-negative, xylem-limited, fastidious bacterium, *Xylella fastidiosa* Wells et al (Fig. 10). A virus was believed to be the pathogen. In 1973, Hopkins et al used electron microscopy to visualize a rickettsia-like bacterium associated with diseased peach tissues. Subsequently, the organism was observed by phase-contrast microscopy and immunofluorescent staining—techniques now used for routine detection. The bacterium was cultured in 1983, and its pathogenic role was confirmed. *X. fastidiosa* is an unusually small bacterium, measuring about 0.25 × 2.5 µm. In grapevines, temporary remission of symptoms has been associated with temperatures of –8 to

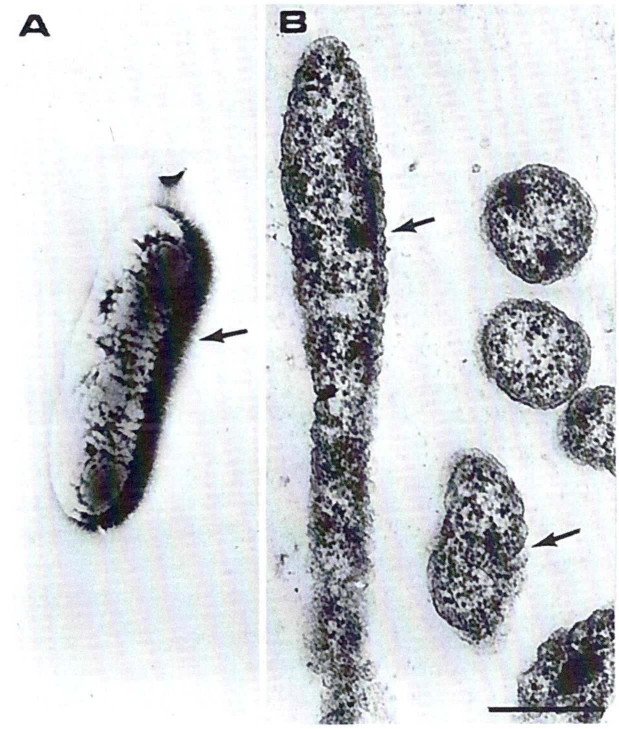

Fig. 10. **A,** Negatively stained cell of *Xylella fastidiosa.* Note the rickettsia-like invaginations of the surface and fimbriae (arrow) on the surface of the cell wall. **B,** Ultrathin section of a cluster of bacteria from the xylem of an infected peach root; bar = 0.5 µm. (Reprinted, by permission, from Wells et al., 1983)

−12°C, suggesting that the bacterium is sensitive to low temperatures. The bacterium is transmitted by leafhopper vectors. It also causes plum leaf scald (PLS) disease and is morphologically indistinguishable from the bacteria that cause Pierce's disease of grapevine (*Vitis vinifera* L.) and leaf scorch diseases of almond (*Prunus amygdalus* Batsch), sycamore (*Platanus occidentalis* L.), oak (*Quercus rubra* L.), elm (*Ulmus americana* L.), mulberry (*Morus rubra* L.), and maple (*Acer rubrum* L.). Some strains, particularly those from grapevine, may be less fastidious in their nutritional requirements than others.

Disease Cycle and Epidemiology

The bacteria enter xylem tissues of peach via the feeding of sharpshooter leafhopper vectors, *Homalodisca coagulata* (Say), *Oncometopia nigricans* (Walker), and possibly other undescribed species. Once in the plant, the bacteria proliferate systemically throughout the xylem. In peach, bacterial numbers are small or zero in leaf and stem tissue but large in root tissue. In plum, bacterial numbers are great in all these tissues. The geographic range of PPD coincides with that of the vectors, which usually overwinter and breed on natural host reservoirs such as wild members of the genus *Prunus* and grasses, especially Johnson grass (*Sorghum halapense* (L.) Pers.). However, in South America, despite the presence of PLS in Japanese plum and an abundance of leafhopper vectors, PPD has not been observed in adjacent peach orchards. The occurrence of the predominant vector, *H. coagulata*, is reduced in peach trees with PPD symptoms. This factor plus the small numbers of bacteria in peach leaves and twigs probably account for the slow or nonexistent secondary spread within peach orchards. However, the presence of less attractive PPD trees in a peach orchard may increase vector feeding on noninfected trees. Seasonal patterns of leafhopper occurrence in peach orchards are influenced by rootstock. Leafhopper vectors were more attracted to scions on a U.S. rootstock, Nemaguard, than to a rootstock from Brazil.

Control

The PPD epidemic of 1915–1929 in the southeastern United States was brought under control by a rigorous program of early detection, tree removal, and eradication of the wild plum *Prunus angustifolia* Marsh. from the vicinity of peach orchards. The outbreak of 1971–1978 was attributed to a series of three warm winters from 1971 to 1974 and to relaxation of the control program. Control is currently based on the same methods developed during these two epidemics, especially early detection and removal of symptomatic trees within the orchard and of wild plums adjacent to peach orchards. Antibiotic treatments of individual symptomatic trees with oxytetracycline or penicillin G halt development of PPD but are not eradicative. This method is expensive and difficult to use in commercial plantings. A more promising approach now being developed is the use of tolerant rootstocks and cultivars.

Selected References

Gould, A. B., French, W. J., Aldrich, J. H., Brodbeck, B. V., Mizell, R. F. III, and Andersen, P. C. 1991. Rootstock influence on occurrence of *Homalodisca coagulata*, peach xylem fluid amino acids, and concentrations of *Xylella fastidiosa*. Plant Dis. 75:767-770.

Hopkins, D. L. 1989. *Xylella fastidiosa*: Xylem-limited bacterial pathogen of plants. Annu. Rev. Phytopathol. 27:271-290.

Wells, J. M., Raju, B. C., Hung, H.-Y., Weisburg, W. G., Mandelco-Paul, L., and Brenner, D. J. 1987. *Xylella fastidiosa* gen. nov., sp. nov: Gram negative, xylem-limited, fastidious plant bacteria related to *Xanthomonas* spp. Int. J. Syst. Bacteriol. 37:136-143.

Wells, J. M., Raju, B. C., and Nyland, G. 1983. Isolation, culture and pathogenicity of the bacterium causing phony disease of peach. Phytopathology 73:859-862.

(Prepared by J. M. Wells)

Plum Leaf Scald

Plum leaf scald (PLS) was reported first in the Paraná River delta in Argentina about 1935 and later in Brazil and Paraguay. In 1977, it was reported in the southeastern United States, where it is especially serious on plum in Alabama, Georgia, and Florida. PLS also has been detected in South Carolina, Texas, and Arizona. The disease causes severe losses on Japanese plum (*Prunus salicina* Lindl.) and also affects other *Prunus* spp., such as *P. domestica, P. cerasifera, P. insititia, P. americana, P. simonii, P. munsoniana,* and *P. triflora.*

Symptoms

Plum leaves first exhibit a slight, irregular chlorosis or browning along the margin or tip in early to midsummer, which intensifies as the season progresses (Plate 83). As the necrotic area spreads from the leaf margin, the brown, dry margin becomes separated from the unaffected area by a diffuse chlorotic band. Trees with generalized PLS have an overall scorched appearance. During the early stages, leaf scald may occur on only a few twigs or individual large branches. In advanced stages of PLS, growth stops, terminals become brittle, and dieback occurs, followed by the death of branches and of the entire tree, often within two to three years after PLS symptoms first appear. Fruit of highly susceptible cultivars fail to develop to normal size, and production is reduced.

Causal Organism and Epidemiology

Bacteria associated with PLS and phony peach disease (PPD) are serologically and morphologically indistinguishable but can be serologically distinguished from the Pierce's disease bacterium. All three of these diseases are caused by the gram-negative, xylem-limited, fastidious bacterium *Xylella fastidiosa* Wells et al. Xylem-limited bacteria, associated with PPD and PLS, have been transmitted from peach trees with PPD symptoms to plum and from plum trees with PLS symptoms to peach by grafting root sections. Disease symptoms become visible within two years after grafting. In Argentina, however, peach trees adjacent to areas with PLS exhibited no PPD symptoms, nor has PPD been reported from South America. Plum leaves, stems, and roots favor bacterial growth, and bacteria may become numerous in these tissues, in contrast to peach, where bacteria are numerous only in root tissue. Peach may be a dead-end host for the pathogen, because insect transmission from peach to peach or peach to plum is low. The bacterium is vectored by sharpshooter leafhoppers such as *Homalodisca coagulata* (Say) and *Oncometopia nigricans* (Walker) and by spittlebugs (family Cercopidae). In the southeastern United States, native and wild plums are considered important reservoirs for PPD and PLS.

Control

Control of PLS involves early detection and removal of symptomatic trees within plum orchards and removal of wild plums adjacent to the orchard. Management of ground cover in and near the orchard is important because some broadleaf and perennial weeds, such as Johnson grass (*Sorghum halapense* (L.) Pers.), serve as alternate hosts for both the bacterium and insect vectors. Although the bacterium is sensitive to oxytetracycline, its use is not considered practical. Plum cultivars vary greatly in susceptibility. Cultivar Santa Rosa is very susceptible; large numbers of bacteria are detected in both twigs and roots. Santa Rita and Sanguinea appear to be more resistant, and Homeside, seedling Methley A-21, Mariposa, and Morris had fewer bacteria and milder symptoms. The more resistant cultivars should be considered when orchards are planned. Where PLS is not endemic, long-range spread can be avoided by use of disease-free rootstocks. New orchards should not be planted near diseased trees or wild plums, and diseased or infected trees should be removed promptly.

Selected References

French, W. J., and Feliciano, A. 1982. Distribution and severity of plum leaf scald disease in Brazil. Plant Dis. 66:515-517.

French, W. J., and Kitajima, E. W. 1978. Occurrence of plum leaf scald in Brazil and Paraguay. Plant Dis. Rep. 62:1035-1038.

Raju, B. C., Wells, J. M., Nyland, G., Brlansky, R. H., and Lowe, S. K. 1982. Plum leaf scald: Isolation, culture, and pathogenicity of the causal agent. Phytopathology 72:1460-1466.

Wells, J. M., Raju, B. C., Thompson, J. M., and Lowe, S. K. 1981. Etiology of phony peach and plum leaf scald diseases. Phytopathology 71:1156-1161.

(Prepared by D. F. Ritchie)

Diseases Caused by Mycoplasmalike Organisms

In 1967, Y. Doi and co-workers proposed that many yellows diseases may be caused by a group of nonculturable, wall-less prokaryotes that are referred to as *mycoplasmalike organisms* (MLOs). Although MLOs are graft-transmissible, they cannot be transmitted by sap inoculations. MLOs reside primarily in phloem tissues and are transmitted by specific phloem-feeding insects, the most important being leafhoppers and psyllas. Once an insect vector acquires an MLO from an infected plant, it remains infective for the rest of its life. Thus, successful management of MLO diseases in an orchard depends on the identification and elimination of pathogen reservoirs and insect vectors.

Cherry Albino

The albino disease of cherry has been found only in the Rogue River Valley of Oregon. Following its appearance in 1937, this disease spread very rapidly and eliminated commercial cherry production for a time. The absence of the disease in more recent cherry plantings in the Medford, Oregon, area suggests that the pathogen may not have survived in the wild *Prunus* species of that region.

Symptoms

Many of the foliar, fruit, and limb symptoms of albino disease are similar to those described for X-disease. The most notable difference between the diseases is the failure to transmit cherry albino to peach trees, a host easily infected with the X-disease MLO. In cherry trees, however, the albino pathogen was apparently more virulent than the X-disease MLO, because trees growing on mazzard rootstocks were killed one to four years after infection; cherry trees on mazzard rootstocks infected with the X-disease MLO survive much longer.

Causal Agent

Electron micrographs showed the presence of MLOs in the phloem of albino-diseased cherry. Because this disease is no longer found, no serological or genetic relationships between the albino MLO and other plant-pathogenic MLOs have been determined.

Selected References

Cameron, H. R. 1976. Albino. Pages 204-205 in: Virus Diseases and Noninfectious Disorders of Stone Fruits in North America. U.S. Dep. Agric. Agric. Handb. 437.

Florance, E. R., and Cameron, H. R. 1978. Three-dimensional structure and morphology of mycoplasmalike bodies associated with albino disease of *Prunus avium*. Phytopathology 68:75-80.

(Prepared by B. C. Kirkpatrick)

Cherry Blossom Anomaly

Blossom anomaly was observed on a Bing sweet cherry tree near Wenatchee, Washington, in 1964; three years later, it was shown to be caused by a graft-transmissible agent. The disease is of minor importance.

Symptoms

Symptoms first appear at bloom time as varying amounts of green along the veins and midribs in the normally white petals. The amount of green coloration varies from narrow streaks to broad bands covering entire petals (Plate 84). In some instances, the entire flower bud assumes a leaflike appearance except for the yellow anthers in the center. In some flowers, a few of the anthers differentiate into wide, white, petal-like structures with green centers. In others, the entire stamen resembles a narrow petal. In a few instances, pistil-like structures develop on tops of the stamens. Fruit that develop from affected blossoms are more pointed than normal and have a persistent style and rough skin (Plate 85).

Infected peach and nectarine trees appear paler (chlorotic) and are less vigorous than healthy trees, but no symptoms have been observed in the blossoms. Symptoms have not been observed on inoculated apricot, plum, apple, or pear trees.

Causal Agent

An MLO was associated with diseased blossoms. Reliable transmission requires use of symptomatic fruit spurs as inoculum sources. An incubation of one to three years is required before symptoms develop. The disease was identified in only a single Bing cherry tree, and repeated surveys of the same orchard for more symptomatic trees were negative. Natural spread of the disease has not occurred among healthy trees near a diseased tree growing in an experimental cherry orchard. Also, repeated attempts to transmit the pathogen from cherry to aster plants via the six-spotted leafhopper (*Macrosteles divisus* (Uhler)) failed.

Selected References

Cheney, P. W., and Parish, C. L. 1969. A graft-transmissible blossom anomaly of cherry. Phytopathology 59:10.

Parish, C. L., Cheney, P. W., and Anthon, E. W. 1976. Blossom anomaly of cherry. Pages 211-212 in: Virus Diseases and Noninfectious Disorders of Stone Fruits in North America. U.S. Dep. Agric. Agric. Handb. 437.

Parish, C. L., and Florance, E. R. 1983. Mycoplasmalike bodies associated with blossom anomaly. 12th Int. Symp. Fruit Tree Virus Dis. Acta Hortic. 130:137.

(Prepared by C. L. Parish)

Peach Red Suture

Red suture of peach was reported from Michigan and Maryland in 1911. In recent years, it has not been economically important. Unconfirmed cases have been noted in southern Indiana (pre-1951) and in western Colorado (1974 and 1980). In the United States, distribution of the disease closely follows that of peach yellows. Peach red suture also has been observed in eastern Canada, Russia, and possibly France and Israel.

Symptoms

Symptoms appear on mature trees 22 months after infection. In contrast, nursery trees propagated from infected buds may not exhibit symptoms in their second year of growth. Peach red suture is most notably characterized by fruit symptoms, but foliage and shoot growth also are affected. Fruit symptoms are similar to those associated with exposure to the herbicides 2,4-D or 2,4,5-T and with peach yellows (also called little peach) disease. Diseased fruit show a roughened, bumpy, swollen, or bulged suture with intense internal color along the swelling (Plate 86). The surface of suture swellings is blotched or splashed with intense color (dark red to purple on red cultivars, a deeper yellow on yellow cultivars). The suture area ripens early and becomes soft, while the opposite side is green and hard; the flesh of this area is coarse, stringy, watery, and easily bruised. Affected fruit often have an insipid flavor. Symptoms on trees infected with red suture occur annually, whereas those caused by herbicides are temporary.

Leaves of infected trees are yellowish green to greenish bronze, especially several weeks after petal fall and just before fruit maturity. Application of nitrogen can improve leaf color to a nearly normal green, but diseased trees often develop autumn coloration prematurely. On vigorous two-year-old branches, new shoots develop into short, spurlike structures with abnormal leaf crowding and clustering along the branch. In contrast, some buds on healthy trees remain dormant, some grow into long shoots, and only a few develop this spurlike growth. Trees infected for several years have open centers due to reduced shoot growth and may survive eight years or longer.

Causal Agent

The causal agent of red suture has not been confirmed. The disease symptoms are similar to those caused by the peach yellows MLO, and as with peach yellows, budwood is cured of disease by immersion in hot water (50°C) for 3 min. However, attempts to transmit the disease with the leafhopper *Macropsis trimaculata* (Fitch) (the vector of the peach yellows MLO) have failed.

Disease Cycle and Epidemiology

The disease agent is transmitted to healthy peach by grafting and spreads naturally in commercial peach orchards. Localized outbreaks occurred from 1920 to 1940 in Michigan. Critical information is lacking, however, on the causal agent and potential vectors. All peach cultivars tested and one cultivar of Japanese plum (cv. Abundance) are susceptible to red suture. Sweet and sour cherries are immune. Abundance plum had lighter-colored foliage and earlier fruit maturity; similar symptoms may be caused by peach yellows. In the Japanese plum cultivar Red June and the plum hybrid Wickson (*Prunus salicina* × *P. simonii*), graft inoculations with red suture resulted in normal-appearing plants, but plum trees became severely diseased when similarly inoculated with peach yellows.

Control

Red suture is controlled by removal of diseased trees. In addition, orchardists should routinely practice use of certified budwood, rootstocks, and finished trees.

Selected References

Cation, D., and Bennett, C. W. 1951. Red suture. Pages 11-13 in: Virus Diseases and Other Disorders with Viruslike Symptoms of Stone Fruits in North America. U.S. Dep. Agric. Agric. Handb. 10.

Klos, E. J. 1976. Red suture. Pages 133-134 in: Virus Diseases and Noninfectious Disorders of Stone Fruits in North America. U.S. Dep. Agric. Agric. Handb. 437.

Kunkel, L. O. 1936. Heat treatment for the cure of yellows and other diseases of peach. Phytopathology 26:809-830.

Richards, B. L., and Cochran, L. C. 1957. Virus and Viruslike Diseases of Stone Fruits in Utah. Bull. 384. Utah State Agric. Coll., Logan. 131 pp.

(Prepared by H. J. Larsen and H. E. Waterworth)

Peach Rosette

Peach rosette was first described by E. F. Smith in 1881. The disease was first reported in Georgia, and Smith believed that peach rosette was probably a southern strain of peach yellows. Although the geographic range of the disease overlaps with that of peach yellows, peach rosette is found primarily in the southeastern United States as far west as Texas. It has not been reported in the western states or outside the United States. Although sporadic outbreaks of the disease still occur, peach rosette is currently of minor importance.

Symptoms

Peach—Some symptoms associated with peach rosette are similar to those of peach yellows, but there are some important differences. Leaves on diseased limbs are of nearly normal size and morphology and become somewhat chlorotic during the growing season. Shoot internodes are very short; thus, the leaves are borne in very tight clusters or brooms (Plate 87). Very few fruit are produced on symptomatic trees, and the few that do develop ripen abnormally and drop prematurely, in contrast to peach yellows. Very few adventitious shoots develop in the interior of the tree canopy. Also, the peach rosette MLO appears to be more virulent than peach yellows, because systemic symptoms and tree death occur more rapidly.

Plum—Japanese plum (*Prunus salicina*) cultivars are susceptible to the peach rosette pathogen. Leaves on infected trees develop chlorosis and often a reddish blush, but the trees do not produce the dense brooms observed in diseased peach trees.

Causal Agent

DNA hybridization studies show that the MLO associated with peach rosette is similar to, but distinct from, the MLO that causes X-disease. Similarity in host range and geographic distribution suggests that peach rosette and peach yellows MLOs may be related but not identical.

Disease Cycle and Epidemiology

The agent was transmitted to healthy peach or Japanese plum trees by grafting diseased bark patches and to periwinkle (*Vinca rosea*), tomato (*Lycopersicon esculentum*), and tobacco (*Nicotiana glutinosa*) through dodder (*Cuscuta campestris*). Although the insect vector of peach rosette is unknown, natural spread occurred from wild plum (*P. angustifolia*) to commercial peach and plum plantings. Other hosts of peach rosette include almond, apricot, flowering almond, Damson plum, western sand cherry, David peach, wild plums (*P. mexicana* and *P. injucunda*), Manchu cherry, mazzard cherry, mahaleb cherry, sour cherry, sand cherry, common chokecherry, and red maple.

Control

Since peach rosette spreads in orchards, diseased trees should be removed. In addition, wild plum appears to be an important endemic reservoir of the pathogen and should not be allowed to grow near commercial orchards.

Selected References

KenKnight, G. 1976. Peach rosette. Pages 73-76 in: Virus Diseases and Noninfectious Disorders of Stone Fruits in North America.

U.S. Dep. Agric. Agric. Handb. 437.

Seliskar, C. E., and Wilson, C. L. 1981. Yellows diseases of trees. Pages 35-96 in: Mycoplasma Diseases of Trees and Shrubs. K. Maramorosch and S. P. Raychaudhuri, eds. Academic Press, New York.

(Prepared by B. C. Kirkpatrick)

Peach Yellows

Epidemic outbreaks of peach yellows (also called little peach) caused significant tree losses during the 19th and early 20th centuries. The disease epidemics occurred primarily on the eastern coast of the United States from Maryland to South Carolina and in some peach-growing districts east of the Mississippi River from Michigan to Tennessee. Peach yellows has not been found in the far western or southern states or outside North America. Currently, the incidence of this lethal disease is low.

Symptoms

Leaf buds on diseased peach trees break dormancy two to four weeks earlier than those on healthy trees. Leaves that develop from these buds are narrower and smaller than healthy leaves, and as the season progresses, they become chlorotic and often develop red spots. Leaves borne on the proximal portions of diseased limbs abscise prematurely, and dense tufts of leaves often develop on the branch terminals. Numerous upright, slender shoots develop (Plate 88) in the interior of the tree canopy. The numerous small leaves borne on these shoots are often severely chlorotic. These symptoms develop first on the scaffold inoculated by an infectious MLO vector. As the disease progresses, the diseased limbs die back from the terminals, and tree death occurs one to five years later.

Fruits produced on diseased limbs develop prematurely, usually two to three weeks ahead of healthy fruit. Although the fruits are of normal or larger than normal size, they taste bland or even bitter. In red-skinned cultivars, the fruit surface develops highly pigmented spots with red streaks in the flesh (Plate 89) and a pronounced red color around the pit.

Causal Agent

DNA hybridization and sequence analyses show that the MLO that causes peach yellows is genetically related to, but distinct from, the X-disease MLO. However, the symptoms produced by the peach yellows MLO are clearly different from those produced by the X-disease MLO. Also, the plum leafhopper (*Macropsis trimaculata* (Fitch)), which transmits peach yellows, does not transmit X-disease.

Disease Cycle and Epidemiology

The peach yellows agent is transmitted by grafting and by the plum leafhopper. Development of disease symptoms requires up to three years in field trees but less than two months under greenhouse conditions. The economic hosts of the pathogen include all cultivars of peach, nectarine, apricot, and almond. All other *Prunus* species that were tested by graft-inoculation were found to be susceptible.

Control

Although peach yellows is currently uncommon, it has the potential to reappear at any time. Historical recommendations for control include removing diseased peach trees as soon as they are found, a suggestion that is supported by the observation that peach supports comparatively high populations of the pathogen. Because Japanese plum (*Prunus salicina*) cultivars are susceptible to infection (see Peach Red Suture) and are the preferred hosts of the leafhopper that transmits peach yellows, plum trees planted near peach orchards in areas where the disease occurs should be carefully monitored and removed if they test positive for the pathogen.

Selected Reference

Pine, T. S., and Gilmer, R. M. 1976. Peach yellows. Pages 91-95 in: Virus Diseases and Noninfectious Disorders of Stone Fruits in North America. U.S. Dep. Agric. Agric. Handb. 437.

(Prepared by B. C. Kirkpatrick)

X-Disease

X-disease is an economically important disease of peach and cherry in most of the eastern and western fruit-growing regions of North America. Although California cherry growers were familiar with a disease they called "cherry buckskin" in the early 1920s, its infectious nature was not established until 1931. Later graft-transmission studies showed that the buckskin pathogen also caused a lethal decline disease of peach, "leaf casting yellows." In 1933 in Connecticut, a peach disorder named X-disease was characterized by symptoms very similar to those of western leaf casting yellows. Subsequent graft-transmission studies, which have been substantiated by serological and nucleic acid hybridization assays, showed that the eastern and western (*Prunus*) strains of the mycoplasma-like organisms were similar, and the term X-disease was adopted to describe this group of "yellows"-inducing *Prunus* pathogens.

Symptoms

Cherry—The pathogen causes similar symptoms in both sweet and sour cherries, but the severity of symptoms varies in different climatic regions. In general, X-disease symptoms are more severe, and the disease progresses more quickly in areas with long, hot growing seasons, such as the Central Valley of California. The productivity of infected cherry trees, however, is reduced in all geographic regions.

X-diseased trees undergo a rapid or slow decline, depending on the rootstock. Infected trees on *Prunus mahaleb* rootstocks decline rapidly, usually following the onset of warm summer temperatures, and usually die within three years after infection. Tree death is rapid because mahaleb rootstocks are immune to infection by the X-disease MLO, and this immunity is mediated by a hypersensitive response that kills mahaleb phloem elements in the region of the graft union after exposure to the pathogen. The ensuing hypersensitive reaction effectively girdles the cherry scion. Because these events occur quickly, such trees usually have normal-sized leaves that exhibit a reddish blush on the leaf undersides and an upward rolling of the leaf margins prior to tree collapse (Plate 90). The fruit on affected trees is usually of normal appearance but may be sweeter than fruit on healthy trees because extra sugars are diverted to fruits following girdling at the graft union. Symptoms on the woody cylinder consist of pits and grooves at the junction of the scion and rootstock (Plate 91), and the mahaleb rootstock may produce numerous suckers.

In contrast to the rapid decline observed on hypersensitive rootstocks, X-disease MLO infection of cherry cultivars grafted on susceptible rootstocks, such as mazzard (*P. avium*), Colt (*P. avium* × *P. pseudocerasus*), or Stockton Morello (*P. cerasus* L. var. *austera* L.), decline slowly over several years. Symptoms appear first in the inoculated limbs or scaffold, whether introduced there by insects or by graft-inoculations. In two or more years, the entire tree becomes systemically infected. Leaves on infected limbs are smaller than normal and develop a red tinge on the upper lamina at mid–growing season. Terminal shoot growth is reduced or absent. Typical fruit symptoms, which are more pronounced in the warmer growing areas, include late or delayed maturation, incomplete color development, conical shape, often shorter-than-normal pedicels (Plate 92), and bland-flavored, watery flesh. As the disease progresses, dieback of limbs and scaffolds is followed by tree death. In the late stages of infection, there is a paucity of feeder roots, and the entire root system eventually dies.

Peach—X-disease symptoms in peach first appear in the scaffold that was inoculated by the infectious insect vector. Typically, the pathogen requires more than two years to completely colonize a mature tree. Leaves on infected trees appear normal until early to mid-June, when irregularly shaped, chlorotic or necrotic lesions develop in the leaf lamina (Plate 93). This necrotic tissue dehisces, resulting in shot-hole symptoms on the leaves. As the season progresses, more lesions develop, affected tissues abscise, and eventually the leaves appear tattered. Older symptomatic leaves are prematurely cast, and new terminal leaves may or may not develop, depending on the vigor of the tree. Fruits borne on infected scaffolds develop later, are less numerous and smaller, and lack flavor. Chronically infected trees exhibit scaffold dieback, and tree death usually occurs after five years.

In California, there is another MLO disease called peach yellow leaf roll. Recent DNA hybridization analyses indicate that this disease can be caused by several genetically distinct MLOs, only one of which is similar to the X-disease MLO. In contrast to X-disease, peach yellow leaf roll causes chlorosis, downward curling, and swelling of the midrib and lateral veins at midsummer (Plate 94) due to the deposition of cork layers along the length of the veins; leaves with necrotic lesions or shot holes are conspicuously absent, and premature fruit drop dramatically reduces fruit yields. Also, the peach yellow leaf roll MLOs are more virulent than typical X-disease strains; mature trees become completely symptomatic within two years after infection and die after three or four years.

Almond, apricot, French prune, and Japanese plum—Recent studies on the epidemiology of X-disease and peach yellow leaf roll in California have identified several other species of *Prunus* that are naturally infected with the X-disease MLO. The impact of MLO infection on these hosts, however, is variable. In almond and apricot, infected trees showed no specific leaf symptoms and no apparent effect on fruit quality or quantity. Similarly, no specific detrimental effects were associated with the infection in naturally infected French prune trees, except for a slight reduction in the amount of terminal shoot growth. In contrast, several cultivars of Japanese plum were severely affected by the X-disease MLO. Foliar symptoms consisted of general chlorosis and reddening, and as the season progressed, all leaves turned red and later became necrotic. By midsummer, high concentrations of the X-disease MLO were detected in the fruit peduncles and leaf petioles. Fruit size, and to a lesser extent fruit quality, was reduced in infected trees. More importantly, many of the infected trees died within one to two years.

Causal Agent

Development of serological and nucleic acid hybridization assays has facilitated differentiation of plant-pathogenic MLOs. These assays show that the eastern and western strains of the X-disease MLO are very similar and are genetically and serologically related to other MLOs that cause American fruit and nut tree diseases (peach yellows, peach rosette, and walnut and pecan witches'-broom). The North American X-disease MLO strains are genetically and serologically unrelated to MLOs found in stone fruit trees growing in Europe.

Surveys have clearly showed that outbreaks of peach yellow leaf roll occur in peach orchards planted near old pear orchards. Commonly, a distinct gradient of diseased peach trees occurred adjacent to pear, which strongly suggested that pear trees were a pathogen reservoir and/or a source of the MLO vector. DNA hybridization assays using probes specific for X-disease MLOs and for other MLO-DNA confirmed that MLOs similar to those that cause peach yellow leaf roll were present in pear trees. Although the exact relationship between the pear-infecting MLOs and those of peach yellow leaf roll requires clarification, these findings indicate that peach and pear culture may not be compatible where these pathogens occur.

Disease Cycle and Epidemiology

In the western United States, the important X-disease MLO vectors are *Fieberiella florii* (Stål), *Colladonus montanus* (Van Duzee), *C. geminatus* (Van Duzee), *Scaphytopius acutus* (Say), and some *Paraphlepsius* and *Osbornellus* species. *Paraphlepsius irroratus* (Say) appears to be the most important MLO vector of X-disease in Michigan and probably in other eastern states.

In order to manage vector populations, it is important to understand their biology. For example, in California *C. montanus* overwinters as adults on sugar beets (*Beta vulgaris* L.) and reproduces on dicotyledonous weeds such as curly dock (*Rumex crispus* L.) and burr clover (*Medicago hispida* Gaertn.). When sugar beets are harvested or orchard weeds dry up in the summer, the vectors migrate into cherry and peach trees. *Paraphlepsius irroratus* reproduces primarily on monocotyledonous weeds but feeds in trees at night. Leafhoppers that have fed on weeds infected by the X-disease MLO, such as burr clover, are likely to become infectious. In orchards that have experienced severe losses to X-disease, growers may benefit by removing resident weeds that are potential pathogen reservoirs and breeding hosts of the vector. *F. florii*, the cherry leafhopper, reproduces on perennial woody hosts, including many orchard trees and ornamental plants such as privet (*Ligustrum* spp.) and boxwood (*Buxus sempervirens* L.). These ornamentals are not considered to be MLO reservoirs but are potential sources of large populations of this leafhopper vector. For this reason, California growers should avoid landscaping their residences with such ornamentals. In areas with high disease incidence, some growers have benefited from one or two postharvest insecticide applications on their orchards.

Control

Plant-pathogenic MLOs are sensitive to oxytetracycline antibiotics. Trunk or scaffold limb injections of oxytetracycline improve tree vigor and partially restore fruit production. But treatments must be repeated annually, and treated trees nevertheless gradually decline. In addition, the costs involved with product registration have precluded legal use of oxytetracycline on several *Prunus* varieties.

The most important, cost-effective, and environmentally sound disease management practice is to identify and remove infected plant sources. In the eastern United States, wild chokecherry (*P. virginiana* L.) is an important reservoir host, and its removal within 200 m of an orchard can significantly reduce the incidence of peach X-disease. In California, MLO-infected chokecherry and bitter cherry (*P. emarginata* (Douglas ex Hook) Walp.) were found throughout a cherry-growing district in the Sierra Nevada but not in the major cherry district located in the Central Valley. In the Central Valley and probably other areas of cherry production, the X-diseased cherry trees are the most important pathogen source. Microscopic, serological, and DNA hybridization analyses show that the pathogen persists at high concentrations in leaves of sweet cherry from July through October, and the leafhopper vector readily acquires the pathogen from cherry trees at this time of year. Pathogen concentrations are highest in infected trees grown on susceptible rootstocks, such as mazzard. Thus, it is essential to remove all infected cherry trees immediately. Peach, in contrast, is a very poor X-disease MLO reservoir. Therefore, immediate removal of diseased peach trees is not as critical as removing infected cherry trees.

In California, the annual rate of newly infected X-diseased trees was compared over a five-year period in 15 orchards that were operated under different orchard management practices: 1) diseased trees were not removed, and insecticides were not applied; 2) diseased trees were removed, but insecticides were not applied; or 3) diseased trees were removed after insecticides were applied. In this experiment, ignoring the disease (practice 1) resulted in much higher disease incidence than management practices 2 or 3.

Selected References

Nyland, G., and Schlocker, A. 1951. Yellow leaf roll of peach. Plant Dis. Rep. 35:33.

Purcell, A. H., Nyland, G., Raju, B. C., and Herbinger, M. R. 1981. Peach yellow leafroll epidemic in northern California: Effects of peach cultivar, tree age, and proximity to peach orchards. Plant Dis. 65:365-368.

Rawlins, T. E., and Horne, W. T. 1931. Buckskin, a destructive graft-infectious disease of the cherry. Phytopathology 21:331-335.

Stoddard, E. M. 1934. Progress report of investigations on a new peach trouble. Conn. Pomol. Soc. Proc. (1933) 43:115-117.

Uyemoto, J. K. 1989. Union aberration of sweet cherry on *Prunus mahaleb* rootstock associated with X-disease. Plant Dis. 73:899-902.

(Prepared by B. C. Kirkpatrick, J. K. Uyemoto, and A. H. Purcell)

European Stone Fruit Yellows

Symptoms of decline in apricot and Japanese plum were first observed in France and Italy in the early 20th century and were given the names apricot chlorotic leaf roll and plum leptonecrosis or plum decline. Apricot chlorotic leaf roll also was observed in Greece, Hungary, Romania, Spain, Switzerland, and the former Yugoslavia; plum leptonecrosis seemed to be present in all European countries where Japanese plum was grown. Both diseases are among the most important infectious disorders of each crop and the predominant causes of decline and death of productive trees. More recently, similar symptoms were described in almond and peach in Germany and Spain. These decline diseases of almond, apricot, peach, and plum are now collectively referred to as European stone fruit yellows (ESFY). ESFY is known to occur in all Mediterranean countries of Europe and as far north as Germany.

Symptoms

Apricot—Symptoms of ESFY are most apparent just before flowering and at the end of the growing season. In spring, leaves emerge on infected trees before flowers. Later in the season, leaves show interveinal yellowing, are conically rolled along the longitudinal axis (Plate 95), and may be brittle and smaller than normal, with a rough, stiff texture. In young trees, these symptoms are generally distributed throughout the tree. In older trees, only a few twigs or scaffolds may be affected initially, with symptoms becoming more generally distributed in later years. In late summer and fall, early budbreak produces a proliferation of weak shoots with short internodes, numerous rudimentary buds (Plate 96), and small, discolored, rolled leaves. These shoots are often killed during a cold winter. Infected trees can be recognized in early winter by delayed leaf fall from the off-season vegetative growth. This late growth, however, has not been observed in Germany.

Diseased trees decline over a period of one to four years following the onset of symptoms. They are characterized by poor growth and shortened internodes. Flowering is usually poor, irregular, and staggered over weeks. Fruits are fewer in number, small, and nonuniform and may drop prematurely. Cuts into affected branches reveal discolored or dead phloem tissue, especially following low winter temperatures. Gummosis and stem pitting have also been reported in declining trees.

The symptoms described are typical for trees grown on plum rootstocks (*Prunus cerasifera*, *P. domestica*, or *P. domestica* L. subsp. *insititia* (L.) C.K. Schneid.), which are not affected by the disease and survive the death of the apricot scion. Apricot trees on peach rootstocks are more severely affected. They show yellowing and rolling of leaves, defoliation in summer or early fall, and death within a few months or during the next winter. Trees on apricot rootstocks respond in an intermediate fashion.

The scion cultivar and undefined environmental factors also influence symptom expression. For instance, some cultivars show no growth abnormality, and others that do not exhibit obvious foliar symptoms die during winter after displaying typical phloem necrosis (Plate 96). Severe frost always damages the sensitive phloem of diseased trees and enhances their decline.

Japanese plum—Symptoms induced in Japanese plums are similar in several respects to those in apricots. The first symptoms, in summer, are smaller, slightly chlorotic, and cylindrically (rather than conically) rolled leaves that become thick and brittle and often brownish red (Plate 97). Defoliation is earlier than normal, and often new growth starts between October and December. In apricot, early emergence of leaves can occur during winter or early spring. Diseased trees are in leaf during bloom and are easily distinguished from healthy trees showing only white flowers. Fewer flowers and fruit develop on diseased trees, and the fruit are smaller and ripen later than on healthy trees. After a winter frost, necrosis can be observed in the phloem and sometimes in the youngest xylem (Plate 98). The decline of the trees is somewhat slower than in apricots. Diseased branches die within a few years, and eventually the entire tree dies. Resistant plum rootstocks will produce vigorous suckers as the scion declines.

Peach—Symptoms vary with the cultivar. In some white-fleshed cultivars, symptoms consist of red foliage in summer or early fall and a slight rolling, curling, and premature casting of leaves. In some yellow-fleshed cultivars, symptoms resemble those of X-disease and peach yellow leaf roll. In these cultivars, leaves appear normal until midsummer. They then become slightly chlorotic or pale green and develop necrotic lesions in the leaf lamina. Later, these lesions abscise to give a "shot-hole" appearance, and the overall leaf yellowing becomes more pronounced. Simultaneously, the leaf margins roll upward longitudinally, leaf tips curl downward, and leaves turn hard and brittle and abscise prematurely (Plate 99). More specific symptoms include swollen midribs resulting from corky depositions and a yellow or red coloration of the enlarged lateral veins. Premature break of leaf buds and phloem discoloration have been observed. Vigor and productivity of infected trees are reduced, scaffold branches exhibit dieback, and trees decline within a few years.

European plum (*P. domestica*)—MLOs affecting the stone fruit species just described have also been graft-transmitted to European plum. In infected cultivars, no obvious symptoms are observed, but vigor of the trees is usually reduced. The diseases of peach, almond, and European plum are less extensively characterized than those of apricot and Japanese plum.

Causal Agent

The ESFY causal agents have been transmitted by grafting. Cross-inoculation experiments showed that inocula of apricot chlorotic leaf roll caused typical symptoms of plum leptonecrosis in Japanese plum and vice versa. Furthermore, inocula from symptomatic apricots or Japanese plums when grafted onto peach and almond induced decline symptoms similar to those in naturally infected trees. The disease of peach has been observed in the same areas as apricot chlorotic leaf roll and plum leptonecrosis and has been described as peach decline, peach yellows, or European peach yellows.

Electron microscopy has confirmed the occurrence of mycoplasmalike organisms (MLOs) in phloem sieve tubes of diseased trees. The MLO associated with apricot chlorotic leaf roll was transmitted by dodder to Madagascar periwinkle

(*Catharanthus roseus* (L.) G. Don.) and observed in thin sections of these plants as well. Also, oxytetracycline injections caused remission of symptoms. Studies concerned with molecular characterization of the 16S ribosomal RNA gene and the MLOs associated with naturally infected apricot, Japanese plum, peach, and almond trees showed nearly identical characterization patterns. Thus, ESFY diseases are caused by the same MLO or genetically very similar MLOs. However, strains differing in virulence, host specificity, and the ability to induce off-season growth have been observed. The ESFY MLO is closely related to an MLO associated with apple proliferation. It is distinctly different from the X-disease MLOs of North America and from other mycoplasmas affecting a great number of woody and herbaceous hosts.

Disease Cycle and Epidemiology

Orchard surveys over many years provide evidence that the pathogen is spreading. In general, diseased trees are not found in new orchards until the fourth or fifth year, when 1–3% of the trees may be affected; after this, the annual increase of diseased trees varies from 1 to 20%. Several reports have cited up to 60% losses of trees after 10–15 years. New infections usually are clustered near sites of infected trees or along the border of the orchard. At present, the vector(s) responsible for the observed orchard spread have not been identified.

Identification

Diseased trees are identified on the basis of visual symptoms of premature budbreak, leaf vein enlargement, leaf coloration and configuration, phloem necrosis, and off-season vegetative growth. However, masking of symptoms by unrelated pathogens or environmental factors may complicate diagnosis of ESFY-diseased trees. In certification or quarantine programs, detection still involves grafting to a sensitive indicator host. Rapid and reliable detection is possible by microscopic examination of phloem tissue sections stained with the DNA fluorochrome 4′-6-diamidino-2-phenylindole (DAPI). Furthermore, dot blot, Southern hybridization analyses, and in vitro amplification of pathogen DNA by polymerase chain reaction provide specific and sensitive detection of the ESFY MLOs. The distribution of the pathogen in infected trees is uneven and is influenced by the host and time of year. Thorough sampling and replication are important for accurate diagnosis.

Control

The use of pathogen-tested scion budwood and rootstocks is essential when planting new orchards. The proper choice of a rootstock may prolong production of infected trees. Currently, the best strategy to minimize disease spread is to reduce the pathogen reservoirs by removing infected trees. Also, alternative hosts should be eliminated as they are identified. The value of insecticide applications to control vectors is not known. In the future, resistant or tolerant scion cultivars and rootstocks may be developed and used.

Selected References

Dosba, F., Lansac, M., Mazy, K., Garnier, M., and Eyquard, J. P. 1991. Incidence of different diseases associated with mycoplasmalike organisms in different species of *Prunus*. Acta Hortic. 283:311-320.

Giunchedi, L., Marani, F., and Credi, R. 1978. Mycoplasmalike bodies associated with plum decline (leptonecrosis). Phytopathol. Mediterr. 17:205-209.

Giunchedi, L., Poggi-Pollini, C., and Credi, R. 1982. Susceptibility of stone fruit trees to the Japanese plum decline causal agent. Acta Hortic. 130:285-291.

Lederer, W., and Seemüller, E. 1992. Demonstration of mycoplasmas in *Prunus* species in Germany. J. Phytopathol. 134:89-96.

Llacer, G., Sànchez-Capuchino, J. A., Forner, J. B., Bono, R., and Casanova, R. 1976. Tetracycline treatments of stone fruit trees affected by apricot chlorotic leaf roll. Acta Hortic. 67:141-148.

Morvan, G. 1977. Apricot chlorotic leaf roll. EPPO Bull. 7:37-55.

Sànchez-Capuchino, J. A., and Forner, J. B. 1975. Vegetative disorders in the Japanese plum trees on Myrobalan rootstock in the province of Valencia (Spain). Acta Hortic. 44:93-97.

Tsalis, D., and Rumbos, I. 1980. New observations on peach decline in Greece and its etiology. Acta Phytopathol. Acad. Sci. Hung. 15:279-283.

(Prepared by E. Seemüller and J. A. Foster)

Molières Disease

Molières disease, originally described as *le deperissement de Molières,* affects sweet cherry and European plum (*Prunus domestica*) and was first observed in 1952 in southwestern France (Département Tarn-et-Garonne). The disease, which was devastating and killed thousands of trees, seems still to be restricted to this area, but is now rare.

The first symptoms in both cherry and plum are slightly chlorotic leaves in summer. The following year, the trees bloom abundantly, but flowers are often malformed and fruit set is poor. The few fruit that do develop have short pedicels or peduncles, remain small, and drop prematurely. Small, deformed leaves, rosetting, and poor lignification of young shoots as well as phloem and bark necroses are also associated with the disease. Young shoots die back, and as the disease progresses, all of the scaffolds die. In new orchards, diseased trees are not usually observed until three to seven years after planting. All cherry cultivars and rootstocks, especially *P. mahaleb*, appear to be susceptible. However, plum cultivars differ in susceptibility.

The disease is graft-transmissible, and there is evidence that the pathogen is spread by an insect. Trees treated with oxytetracycline recovered and regained full productivity for up to two years. Fluorescent particles typical for mycoplasmalike organisms (MLOs) have been observed in phloem sieve tubes of diseased plum and cherry trees after staining microtome sections with the DNA-specific fluorochrome 4′-6-diamidino-2-phenylindole (DAPI). MLOs have also been detected in plum by electron microscopy. These findings strongly support an MLO etiology for the disease. The relationship of the agent to other MLOs, especially to the European stone fruit yellows MLOs, has not been elucidated.

Selected References

Bernhard, R., Marénaud, C., Eymet, J., Séchet, J., Fos, A., and Moutous, G. 1977. Une maladie de certains *Prunus*: "Le deperissement de Molières." C. R. Seances Acad. Agric. Fr. 2:178-189.

Dosba, F., Lansasc, M., Mazy, K., Garnier, M., and Eyquard, J. P. 1991. Incidence of different diseases associated with mycoplasmalike organisms in different species of *Prunus*. Acta Hortic. 283:311-320.

(Prepared by E. Seemüller)

Diseases Caused by Plant-Parasitic Nematodes

Plant-parasitic nematodes are microscopic soil-inhabiting invertebrates belonging to the animal phylum Nematoda. The phylum consists of two classes: Adenophorea (single-cell excretory system and no phasmids) and Secernentea (tubular excretory system and phasmids present), which are further divided into about 15 orders. All plant-parasitic nematodes are classified in the orders Dorylaimida and Tylenchida. All plant-parasitic nematodes possess a stylet in the anterior region of the digestive system, through which they parasitize plant tissue by mechanical stylet penetration of host cells and various chemical mechanisms. Members of the order Tylenchida have a stylet with basal knobs (stomatostyle) and a multipart esophagus with a median bulb containing a valve (tylenchoid esophagus). In contrast, the Dorylaimida have stylets that lack basal knobs (odontostyles or onchiostyles) and a bottle-shaped esophagus without a median bulb (dorylaimid esophagus). Nematodes that parasitize stone fruit trees feed on root tissue as ectoparasites, migratory endoparasites, or sedentary endoparasites. As pathogens, nematodes cause infectious diseases of stone fruit by damaging root tissues and disrupting water and nutrient uptake. Plant-parasitic nematodes also predispose trees to other infectious diseases and to environmental stresses caused by physical factors such as cold temperature. Although plant-parasitic nematodes differ in biology and morphology, all have egg, juvenile, and adult stages.

Most published research on plant-parasitic nematodes associated with stone fruit is concerned with peach; however, nematodes also attack almond, apricot, nectarine, plum, French prune, sweet cherry, and tart cherry. At least 19 genera of nematodes have been reported to be associated with stone fruit. Those known to be most injurious include species of root-knot (*Meloidogyne*), root-lesion (*Pratylenchus*), ring (*Criconemella*), and dagger (*Xiphinema*) nematodes.

Nematode management in stone fruit production requires a combination of biological monitoring, obtaining nematode- and virus-free nursery stock, developing nematode-resistant rootstocks, use of nematicide, fallowing, and planting of selective ground covers. Several pre- and postplant nematicides are available, although specific recommendations depend on host, nematode, and regional needs. Integrated nematode management tactics vary among regions and with prevailing conditions.

Root-Knot Nematodes (*Meloidogyne* spp.)

Symptoms

Root-knot nematodes are highly specialized plant parasites, and their infection is characterized by gall formation resulting from hypertrophy. At the cellular level, they feed at sites known as syncytia, which are metabolically very active. As infection progresses, the size of the galls and degree of root deformation increase (Plate 100). Gall size, however, varies with the nematode species. Root-knot nematodes disrupt normal root growth and nutrient and water uptake. The combined effect of these changes is an altered host physiology, reduced root and shoot growth, small shoot and root biomass, leaf chlorosis, low fruit yield, poor fruit quality, and sometimes tree death.

Causal Organisms

Root-knot nematodes are sedentary endoparasites, classified in the order Tylenchida. They possess a tylenchoid esophagus, which is common to many plant-parasitic nematodes, exhibit sexual dimorphism, and have a wide host range. Root-knot nematodes were first observed associated with peach in 1885 in Europe and in 1889 in the United States. At least four species have been reported to be important pathogens of *Prunus* spp. In warm climates and sandy soils, these include *Meloidogyne incognita* (Kofoid & White) Chitwood, *M. javanica* (Trueb) Chitwood, and occasionally *M. arenaria* (Neal) Chitwood. *M. hapla* Chitwood can be a significant pathogen of stone fruit in cooler climates.

Life Cycle and Epidemiology

Like most plant-parasitic nematodes, root-knot nematodes have an egg stage, four juvenile stages, and adults. The first molt occurs in the egg; second-stage juveniles hatch from eggs and penetrate root tissue just behind the root tip (Plate 101). The nematode migrates to the vascular parenchyma tissue and forms syncytia, called giant cells. Once syncytia are formed, the nematode feeds as a sedentary endoparasite and enlarges into a sausage stage. The nematode proceeds through three molts before becoming a saccate (saclike) adult female (Plate 102). Preadult males undergo a complete metamorphosis within the fourth-stage cuticle and emerge as migratory, vermiform males. The female deposits 200 to more than 1,000 eggs in a gelatinous matrix outside the posterior region of her body. The life cycle of a root-knot nematode requires four weeks optimally, or longer if the temperature and host environment are not favorable for nematode development.

Control

Rootstocks with resistance to root-knot nematodes are common in peach production (Plate 103). Nemaguard, Shalil, S-37, Okinawa, Higama, Flordaguard, and Nemared are some of the peach rootstocks with resistance to specific root-knot nematode species and races. Marianna 2624 and Myrobalan 29C plum also have resistance to specific root-knot nematodes. However, rootstocks resistant to root-knot nematodes may be susceptible to other nematodes; for example, the ring nematode, *Criconemella xenoplax*. Therefore, regional differences and horticultural requirements may limit commercial use of resistant rootstocks.

Precautions must be taken not to introduce infected plant material into a new orchard and to remove old roots in a replant situation. Fumigant and nonfumigant nematicides have been used for root-knot nematode control. Methyl bromide, 1,3-dichloropropene, and fenamiphos are the nematicides most commonly used. Soil fumigants are applied preplant on a broadcast, row, or tree site basis. Fumigants are phytotoxic, and hence treated soil must be aerated before planting. Such treatments are effective for only two or three years, and postplanting treatments may be needed. Fenamiphos is the only nematicide registered for postplanting use on stone fruits in the United States. Recommendations for nematicide use vary widely by locality and region.

Selected References

Dropkin, V. H. 1989. Introduction to Plant Nematology, 2nd ed. John Wiley & Sons, New York. 304 pp.

McKenry, M. V. 1985. Nematodes. Pages 127-133 in: Integrated Pest Management for Almonds. Publ. 3308. University of California, Division of Agriculture and Natural Resources, Oakland.

McKenry, M. V. 1989. Nematodes. Pages 139-147 in: Peaches, Plums, and Nectarines: Growing and Handling for Fresh Market. Publ. 3331. University of California, Division of Agriculture and Natural Resources, Oakland.

Rom, R. C. 1983. The peach rootstock situation: An international perspective. Fruit Var. J. 37:3-14.

(Prepared by G. W. Bird and H. Melakeberhan)

Root-Lesion Nematodes (*Pratylenchus* spp.)

Symptoms

Much of the damage caused by root-lesion nematodes results from their ability to induce extensive root necrosis. These nematodes generally feed in the root cortex (Fig. 11). Nematode invasion of young roots induces reddish brown, elongated lesions. Root lesions darken with age, necrosis often spreads to affect the entire root, growth ceases, and the root dies. Severe infection may destroy the fine roots. Heavy infections often result in small root systems and stunted trees. Symptoms are more common on young trees than on old trees.

Causal Organisms

Root-lesion nematodes are migratory endoparasites in the order Tylenchida. They possess a tylenchoid esophagus, and all free-living stages are vermiform and mobile. They are more numerous and injurious in sandy than in fine-textured soils. At least five species of the genus *Pratylenchus* have been reported as pathogens of *Prunus* spp. *Pratylenchus vulnus* Allen & Jensen and *P. penetrans* (Cobb) Filipjev, Schuurmans & Stekhoven are the most damaging species to stone fruit. *P. vulnus* is a major pathogen in warm climates, such as in California, whereas *P. penetrans* is important in cool climates, such as in the northeastern United States and southern Canada.

Life Cycle and Epidemiology

After the first molt inside the egg, the second-stage juvenile hatches. The adult emerges after three additional molts. All free-living stages (second-stage juveniles [J-2] to adults) are infective. Most root-lesion nematode species reproduce sexually, and eggs are deposited in soil or root tissue (Fig. 11). The life cycle under favorable conditions requires about four weeks. The reproductive potential of this genus is high, and in California a preplant population of a single *P. vulnus* per kilogram of soil is considered a risk to a new orchard. Population densities of *P. penetrans* as high as 1,100 per gram of fresh root have been extracted from declining Michigan sweet cherry orchards on mazzard rootstock where bacterial canker caused by *Pseudomonas syringae* pv. *syringae* van Hall or *P. syringae* pv. *morsprunorum* Wormald was also present. Root-lesion nematodes may predispose *Prunus* species to cold injury and to other pathogens.

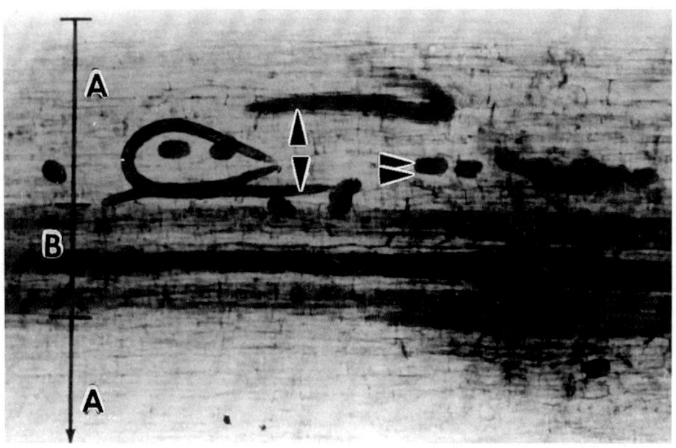

Fig. 11. Migratory feeding (arrowhead) and egg-laying (double arrowhead) behavior of root-lesion nematode (*Pratylenchus* sp.). Nonvascular **(A)** and vascular **(B)** tissues are indicated. (Reprinted, by permission, from Parker et al., 1966)

Control

Dissemination of root-lesion nematodes through infected plant material is common. Thus, utmost care must be taken to produce and plant nematode-free nursery stock. Studies in California have shown that *Prunus japonica* L. is resistant to *P. vulnus*, but relatively little is known about resistance to root-lesion nematodes in stone fruits. Five almond rootstocks, including Garriques and Atocha, are equally susceptible to *P. vulnus,* whereas plum rootstocks Marianna 2624, Myrobalan 29C, Royal apricot, and sand cherry are tolerant. Tzim Pee Tao, Rutgers Red Leaf, and two progenies derived from these rootstocks exhibit partial resistance to *P. penetrans*. A peach cross (Okinawa × Nemaguard) also exhibited resistance to *P. penetrans* and *Meloidogyne*.

Fumigant nematicides are commonly used as preplant treatments. However, reinfestation of fumigated sites develops quickly, often within two years. Even so, the fumigants are helpful in the establishment of young trees in soils where lesion nematodes thrive.

Selected References

Lownsbery, B. F. 1981. Nematode parasites of prunes. Pages 104-107 in: Prune Orchard Management. Publ. 3269. University of California, Division of Agriculture and Natural Resources, Oakland.

McKenry, M. V. 1989. Nematodes. Pages 139-147 in: Peaches, Plums, and Nectarines: Growing and Handling for Fresh Market. Publ. 3331. University of California, Division of Agriculture and Natural Resources, Oakland.

Parker, K. G., Mai, W. F., Oberly, G. H., Brase, K. D., and Hickey, K. D. 1966. Combating replant problems in orchards. Cornell Univ. Ext. Bull. 1169. 19 pp.

(Prepared by G. W. Bird and H. Melakeberhan)

Ring Nematodes (*Criconema* and *Criconemella* spp.)

Symptoms

Ring nematodes are primary pathogens, and they also predispose trees to other pathogens and environmental stresses. Microscopic symptoms of ring nematode-infected trees are similar to those caused by other nematodes; the most prominent symptoms are the destruction and death of feeder roots

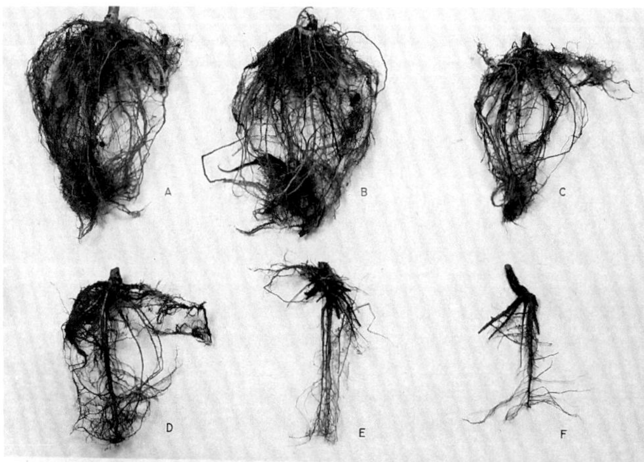

Fig. 12. Peach roots from soil infested with *Criconemella xenoplax* fumigated **(A–C)** or not fumigated **(D–F)** for nematode control. Note stunted growth and sparse appearance of fine roots. (Courtesy A. P. Nyczepir)

(Fig. 12). Fine roots that remain have gray necrosis of the cortex and sometimes fasciculation (dense clusters of fine roots). Feeding by large numbers of nematodes stunts roots and shoots. Infected trees have a thin canopy, and young shoots tend to be slender and willowy. Shoot elongation terminates before the end of the growing season, and infected trees show drought stress earlier than do trees in noninfested soil.

Causal Organisms

Ring nematodes feed as ectoparasites (Fig. 13) and are classified in the order Tylenchida. Distinctive features of ring nematodes are a characteristic body shape and heavy cuticular annulation. Several species of ring nematodes are common in stone fruit orchards, but *Criconemella xenoplax* (Raski) Luc & Raski appears to be the most important species, especially in its role as a predisposing agent to bacterial canker. Predisposition of peach and French prune trees to bacterial canker and to Leucostoma canker caused by *Leucostoma* spp. has been attributed to *C. xenoplax*. In the peach tree short life (PTSL) syndrome in the southeastern United States, *C. xenoplax* is a factor associated with death of peach trees. *C. xenoplax* also frequently occurs in declining Michigan sweet cherry orchards on mazzard rootstocks that are infected by *P. syringae* pv. *syringae* and *P. syringae* pv. *morsprunorum*.

Life Cycle and Epidemiology

Criconemella spp. are assumed to be ectoparasites that induce destructive changes at the feeding sites (Fig. 13). Recent findings, however, show that *C. xenoplax* induces elaborate cellular modifications while feeding as an ectoparasite. Nematodes can feed up to eight days on a single feeding site. All free-living stages (J-2 to adult) feed. Duration of the life cycle is influenced by soil temperature, soil type and moisture level, soil pH, and host species and cultivar. The reported duration of the life cycle ranges from four to eight weeks, but it is more accurately determined when based on degree-days (base 9°C). For example, at 24°C, the life cycle requires 30 days, which is equal to approximately 443 degree-days, whereas at 20°C, 40 days are required. *C. xenoplax* reproduces well at temperatures between 22 and 26°C, but populations also increase during winter when soil temperatures range from 7 to 13°C.

Control

Control of ring nematodes has been accomplished largely through soil treatment with chemical nematicides, especially treatment with 1,3-dichloropropene before planting and fenamiphos after planting. No resistance to *C. xenoplax* has been found in peach rootstocks, but a rootstock candidate identified in South Carolina and Georgia as tolerant to high nematode populations is undergoing extensive field testing. If successful, this rootstock (as yet unnamed) may have tremendous value in alleviating the severe losses often attributed to ring nematodes.

For further discussion of ring nematode control on peaches, refer to the section on peach tree short life. Rotation of fruit crops with wheat appears to be beneficial for reducing ring nematode populations.

Selected References

Brittain, J. A., and Miller, R. W. 1976. Managing peach tree short life in South Carolina. Clemson Univ. Ext. Serv. Bull. 568. 19 pp.

Hussey, R. S., Mims, C. W., and Westcott, S. W. III. 1992. Ultrastructure of root cortical cells parasitized by the ring nematode *Criconemella xenoplax*. Protoplasma 167:55-64.

Nyczepir, A. P. 1991. Nematode management strategies in stone fruits in the United States. J. Nematol. 23:334-341.

Nyczepir, A. P., Zehr, E. I., Lewis, S. A., and Harshman, D. C. 1983. Short life of peach trees induced by *Criconemella xenoplax*. Plant Dis. 67:507-508.

Seshadri, A. R. 1964. Investigation of the biology and life cycle of *Criconemoides xenoplax* Raski, 1952 (Nematoda: Criconematidae). Nematologica 10:540-562.

Westcott, S. W., and Burrows, P. M. 1991. Degree-day models for predicting egg hatch and population increase of *Criconemella xenoplax*. J. Nematol. 23:386-392.

(Prepared by G. W. Bird and H. Melakeberhan)

Fig. 13. Juvenile (left) and adult female (right) of *Criconemella xenoplax* feeding on a plant root. (Courtesy S. W. Westcott III)

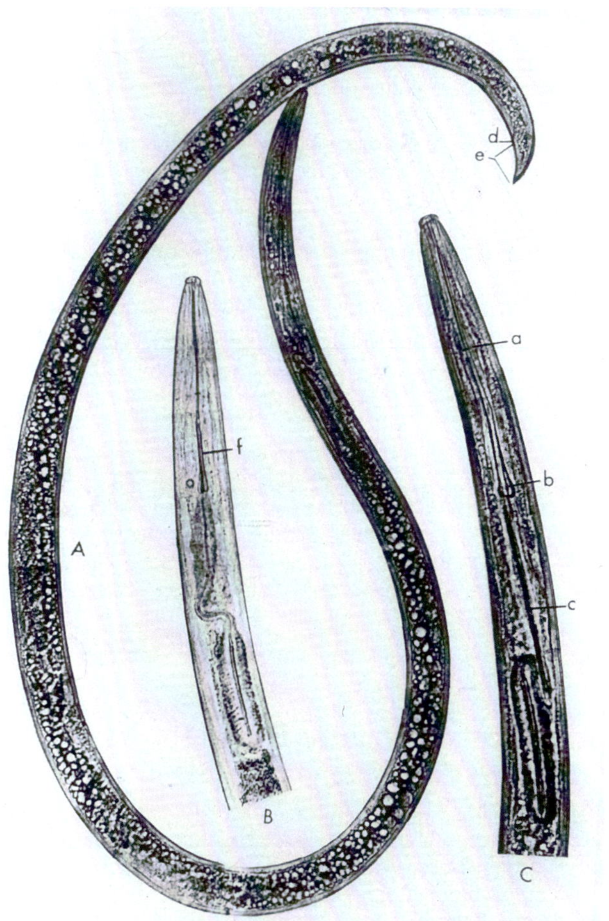

Fig. 14. Female dagger nematode (*Xiphinema americanum*). **A,** Mature female, showing the anus (d) and tail (e) (×128). **B,** Anterior portion of a female, showing the stylet extension (f) (×175). **C,** Anterior portion of a female, showing the guiding ring (a), flanges (b), and developing stylet (c) (×300). (Reprinted from Mai and Lyon, 1975. Used by permission of Cornell University Press)

Dagger Nematodes (*Xiphinema* spp.)

Symptoms

High populations of dagger nematodes injure stone fruit trees, but their major importance is as vectors of certain viruses, such as tomato ringspot virus (TmRSV). Dagger nematodes cause extensive root necrosis and swollen tips of fine root hairs. Under heavy infestations, poor yields and stunted growth are common.

Causal Organisms

Dagger nematodes feed as ectoparasites and are classified in the genus *Xiphinema* in the order Dorylaimida. They possess a dorylaimoid esophagus and a long, slender stylet with flanges at the base, to which muscles are attached (Fig. 14). The stylet resembles a dagger, from which the name is derived. Some species of dagger nematodes have a wide host range and worldwide distribution, and several species are common in stone fruit orchards. The most common dagger nematodes associated with stone fruits are in the *Xiphinema americanum* (Cobb) group, and vector viruses.

Life Cycle and Epidemiology

Dagger nematodes develop through juvenile and adult stages. All stages are vermiform (wormlike) and feed on roots. Depending on the host and environmental conditions, dagger nematodes generally complete one generation per year. Individual nematodes may be able to transmit virus throughout their lives, but viruses are not retained through molts between stages and are not passed from female to egg. Thus, potential for virus transmission declines rapidly unless the nematode population reacquires the virus by feeding on infected host plants.

Feeding by a single viruliferous *X. americanum* can initiate a virus infection. TmRSV, which is vectored by *X. americanum,* causes many severe diseases of stone fruits and can kill peach, cherry, and plum trees before they reach eight years of age.

Control

Fumigant and nonfumigant nematicides are effective against dagger nematodes, but they often are difficult to apply in soils where these nematodes thrive. Long rotations to nonhosts may be necessary to eliminate viruliferous nematodes where TmRSV infestations have been found. Also, other hosts for nematode-transmitted viruses, such as dandelion (*Taraxacum officinale* Wigg.) for TmRSV, must be controlled if virus transmission is to be addressed effectively.

Selected References

Forer, L. B., Powell, C. A., and Stouffer, R. F. 1984. Transmission of tomato ringspot virus to apple rootstock cuttings and to cherry and peach seedlings by *Xiphinema rivesi.* Plant Dis. 68:1052-1054.

Halbrendt, J. M., and Brown, D. J. F. 1992. Morphometric evidence for three juvenile stages in some species of *Xiphinema americanum* sensu lato. J. Nematol. 24:305-309.

Lamberti, F., Taylor, C. E., and Seinhorst, J. W. 1975. Nematode Vectors of Plant Viruses. Plenum Press, New York. 460 pp.

Lownsbery, B. F. 1981. Nematode parasites of prunes. Pages 104-107 in: Prune Orchard Management. Publ. 3269. University of California, Division of Agriculture and Natural Resources, Oakland.

Mai, W. F., and Lyon, H. H. 1975. Pictorial Key to the Genera of Plant Parasitic Nematodes, 4th ed. Cornell University Press, Ithaca, NY.

(Prepared by G. W. Bird and H. Melakeberhan)

Other Nematodes

Pin (*Paratylenchus* spp.), stunt (*Tylenchorhynchus* spp., *Quinisulcius* spp., and *Merlinius* spp.), spiral (*Helicotylenchus dihystera* Cobb), and lance (*Scutellonema* spp.) nematodes have been found in stone fruit orchards. In some cases, they are associated with declining stone fruit trees; however, there are no experimental data supporting their role as pathogens of specific diseases in stone fruits.

Selected Reference

McKenry, M. V. 1989. Nematodes. Pages 139-147 in: Peaches, Plums, and Nectarines: Growing and Handling for Fresh Market. Publ. 3331. University of California, Division of Agriculture and Natural Resources, Oakland.

(Prepared by G. W. Bird and H. Melakeberhan)

Diseases Caused by Viruses and Viruslike Agents

Viruses Spread in Pollen

Prunus Necrotic Ringspot Virus

The first virus-induced Prunus necrotic ringspot disease was described in peach in 1931. Since then, several line pattern and ringspot diseases of *Prunus* have been described on the basis of symptomatology. These include almond calico, apricot line pattern, cherry necrotic ringspot, cherry rugose mosaic, Prunus ringspot, plum line pattern, Stecklenberger, stone fruit ringspot, sour cherry necrotic ringspot, and tatter leaf of peach or cherry. All of these names were used to designate causal agents whose properties were largely unknown. All of these diseases now appear to be caused by variants of Prunus necrotic ringspot virus (PNRSV).

Symptoms

Many variants of PNRSV can be found in most cultivated *Prunus* species. Some variants do not induce symptoms and are detected only by inoculations to woody indicator plants such as Shirofugen flowering cherry (*P. serrulata* Lindl.) or by serological tests such as enzyme-linked immunosorbent assay (ELISA). Other PNRSV isolates produce necrotic spots and shot holes (Plate 104) on young leaves during the first year of systemic infection but few, if any, symptoms on leaves or fruit in later years. Still other isolates produce necrotic reactions in the first year, followed by chronic chlorotic leaf mottle and necrosis, leaf enation, deformity, delayed fruit maturity (Plate 105), and fruit-marking symptoms (Plate 106). A few isolates cause chlorotic line patterns and mosaics on leaves of several plum cultivars. Some PNRSV strains cause calico leaf symptoms on cherry (Plate 107) and almond. Isolates of apple mosaic virus (ApMV), which is serologically related to PNRSV, also cause line patterns and mosaics on plum and almond cultivars. In most *Prunus* species, symptoms expressed depend on the cultivar and the virus isolate.

Causal Agent

PNRSV includes a diverse and complex group of interrelated virus isolates. The virus name is used both as a general term to identify all serologically interrelated members of the ilarvirus subgroup III (which includes viruses that cause mosaic and line pattern diseases of apple, rose, and *Prunus*) and to specifically identify viruses that cause ringspot diseases in several *Prunus* species. Particle shape ranges from isometric to short bacilliform, with axial ratios between 1.0 and 1.5. For most isolates, three sizes of virion particles are found. Also, some isolates have bacilliform particles up to 70 nm long. The genome of PNRSV consists of four molecules of positive-sense, single-stranded RNA of between 0.3 and 1.3×10^6 Da. The virus is relatively unstable in tissue extracts but can be purified from infected cucumber (*Cucumis sativus* L.) cotyledons or from leaves of *Chenopodium quinoa* Willd. by adding antioxidants and reducing agents to the buffer in which the plant material is macerated.

Disease Cycle and Epidemiology

In many *Prunus* species, PNRSV is carried on and in pollen grains and can infect seeds. Because it is transmitted through seed and is easily disseminated by routine plant propagation methods, PNRSV is found wherever cultivated *Prunus* is grown. All *Prunus* species and cultivars appear to be susceptible to one or more isolates. Isolates classified serologically or biologically as PNRSV are also found naturally in rose (*Rosa* spp.) and hop (*Humulus americanus* Nutt.). Isolates classified as ApMV are found naturally in almond (*P. dulcis* (Mill.) D. Webb), apple (*Malus sylvestris* Mill.), birch (*Betula* spp.), hop, hazelnut (*Corylus* spp.), horse chestnut (*Aesculus* spp.), plum (*P. cerasifera* Ehrh.), rose, and raspberry and blackberry (*Rubus* spp.). Isolates of PNRSV and ApMV have been transmitted experimentally to a wide range of herbaceous plants. However, there are no reports of natural occurrence of PNRSV in herbaceous plants other than hop.

Experimental evidence combined with the observed spatial patterns of natural virus spread among orchard trees of cherry and peach following a bloom period indicated that PNRSV infects healthy trees during the pollination process. However, other studies have found that infected seeds can be produced on healthy trees that later do not become infected, suggesting that the mechanism of horizontal transmission is unclear. Recently, PNRSV and prune dwarf virus were transmitted to cucumber seedlings via virus-infected pollen sources and *Thrips tabaci* Lindeman or *Frankliniella occidentalis* (Pergande). No transmissions occurred when pollen and thrips were used separately. The overall importance of thrips control in disease dissemination and tree-to-tree virus spread under orchard conditions, however, is unknown.

Control

Two basic methods are used to control PNRSV-induced diseases: certification programs that provide nurseries with virus-tested selections of important cultivars, and orchard management practices designed to prevent or contain tree-to-tree spread. In general, effective management requires prompt removal of diseased plants, avoiding use of commercial beehives that were used in stone fruit orchards during the preceding two weeks, and use of virus-free wood for any topworking. The disease also has been contained through cross-protection by inoculating surrounding trees with a symptomless virus isolate, although this method of control is not routinely practiced.

Prune Dwarf Virus

Prune dwarf virus (PDV) is associated with several economically important diseases of stone fruits, including yellows of sour cherry, blind wood and narrow leaf of sweet cherry, and dwarf of Italian prune, and together with Prunus necrotic ringspot virus (PNRSV) causes stunt of peach. PDV is similar to PNRSV in causing necrotic spots and shot holes in some *Prunus* species. PDV occurs wherever stone fruits are grown.

Symptoms

Symptoms of sour cherry yellows (Plate 108) consist of chlorotic rings or mottle on young expanding leaves of newly infected sour cherry trees. When tissues in some chlorotic areas become necrotic, they may fall out, causing a shot-hole appearance. Symptoms become progressively milder on young leaves that expand later in the season. These symptoms rarely recur after the first year of infection. One or more years after acute symptoms appear, leaves on infected trees develop a distinctly yellow and green mottle and begin to drop. Leaf casting appears to be accelerated by cool nighttime temperatures (10–16°C) alternating with high daytime temperatures (30–35°C). On chronically infected trees, a reduced number of fruit spurs gives a bare wood appearance (Plate 109). Although yields may be reduced up to 50%, the remaining fruit are large and firm. In sweet cherry, initial symptoms consist of chlorotic spots or rings and some shot holes. Numbers of fruiting spurs are reduced in later years. At least one strain induces narrow leaves that have rough surfaces (Plate 110). Fruit on affected trees tends to be larger and firmer than fruit on healthy trees.

Infected plum and prune develop narrow, straplike leaves that are thicker than normal. Internodes tend to be rosetted in early spring but resume normal expansion later in the season. PDV appears to produce only mild stunting in most peach cultivars and essentially no leaf symptoms. In the Muir peach cultivar, however, PDV infections produce a dense canopy (Plate 111) due to shortening of shoot internodes. Also, when young trees (several peach cultivars were tested) are infected by PDV and PNRSV, growth is severely retarded (peach stunt disease) during the cool spring months, and yields are reduced. Stunted trees have no characteristic leaf symptoms.

Causal Agent

PDV belongs to ilarvirus subgroup IV and has no serological homology with PNRSV isolates. Particles are isometric (19–20 nm in diameter) to short bacilliform up to 73 nm long. Five or more particle classes have been described. Virus particles contain RNA that is presumably single-stranded. The virus can be purified readily from infected cucumber or squash tissue and remains stable for years in the presence of ethylenediamine tetraacetic acid (EDTA).

Disease Cycle and Epidemiology

PDV infects a wide range of *Prunus* species and can be transmitted by sap-inoculations to a number of herbaceous plant species. *Momordica balsamina* L. and *Cyamopsis tetragonoloba* (L.) Taub. appear to differentiate PDV from PNRSV isolates. Even though they are unrelated serologically, PNRSV and PDV isolates share similar biological properties, such as transmission through pollen and seed and dissemination by vegetative propagation. Natural orchard spread of PDV during bloom and virus transmissions to cucumber plants by thrips have also been reported.

Control

Virus certification programs provide nurseries with stone fruit cultivars free of PDV, but no effective strategies have been developed to prevent introduction of the virus into virus-free orchards or its subsequent spread.

Plum Line Pattern Ilarviruses

Plums with foliar line pattern symptoms have been described in every European country and in the United States. At least three serologically distinct viruses have been transmitted mechanically from line pattern–diseased plum cultivars. However, no attempt has yet been made to back-transmit these viruses to the same plum cultivars and compare symptoms under standard conditions. Consequently, although line pattern is internationally recognized as a disease of plums and some other *Prunus* spp., the causal agents of specific diseases are uncertain.

Symptoms

Specific symptoms vary with the virus, cultivar, and perhaps the environment. In general, leaves develop chlorotic lines, bands, or oak-leaf patterns. Veinbanding (Plate 112) and small chlorotic rings can occur. In some cultivars, the patterns appear yellow or yellow-green in the spring and become off-white during the summer. Some cultivars are infected without apparent symptoms.

Causal Agents

Three serologically distinct viruses have been associated with plum line pattern diseases: Danish plum line pattern strain of Prunus necrotic ringspot virus, European plum line pattern strain of apple mosaic virus, and a serologically distinct virus designated American plum line pattern virus. Danish and European line pattern viruses are serologically related but not identical. All three viruses belong to the ilarvirus group and thus are multicomponent with tripartite genomic RNA (i.e., three molecules make up the complete virus).

Disease Cycle and Epidemiology

Early descriptions of line pattern viruses did not include serology to distinguish among isolates; therefore, host ranges specific for the three viruses are uncertain. Line pattern symptoms have been induced after graft-inoculation on a wide range of plum, peach, and cherry cultivars and on other *Prunus* species. Although many ilarviruses are transmitted through pollen and seed, grafting is the only known mode of transmission for these three viruses.

Control

Virus certification programs provide a source of trees free of plum line pattern diseases. However, in practice most growers retain infected trees until they are no longer profitable, because most, if not all, plum line pattern diseases do not greatly affect fruit production or quality.

Selected References

Fulton, R. W. 1981. Ilarviruses. Pages 377-413 in: Handbook of Plant Virus Infections and Comparative Diagnosis. E. Kurstak, ed. Elsevier/North-Holland Biomedical Press, Amsterdam.

Fulton, R. W. 1983. Ilarvirus group. Descriptions of Plant Viruses, No. 275. Commonwealth Mycological Institute and Association of Applied Biologists, Kew, Surrey, England. 3 pp.

George, J. A., and Davidson, T. R. 1963. Pollen transmission of necrotic ringspot virus and sour cherry yellows viruses from tree to tree. Can. J. Plant. Sci. 43:276-288.

Gilmer, R. M., Nyland, G., and Moore, J. D. 1976. Prune dwarf. Pages 179-190 in: Virus Diseases and Noninfectious Disorders of Stone Fruits in North America. U.S. Dep. Agric. Agric. Handb. 437.

Greber, R. S., Teakle, D. S., and Mink, G. I. 1992. Thrips-facilitated transmission of prune dwarf and Prunus necrotic ringspot viruses from cherry pollen to cucumber. Plant Dis. 76:1039-1041.

Kirkpatrick, H. C., and Fulton, R. W. 1976. Virus diseases of plum and prune plum line pattern. Pages 166-175 in: Virus Diseases and Noninfectious Disorders of Stone Fruits in North America. U.S. Dep. Agric. Agric. Handb. 437.

Mink, G. I. 1991. Prunus necrotic ringspot virus. Pages 335-356 in: Plant Diseases of International Importance, Vol. 3: Diseases of Fruit Crops. J. Kumar, H. S. Chaube, U. S. Singh, and A. N. Mukhopadhyay, eds. Prentice Hall, Englewood Cliffs, NJ.

Nemeth, M. 1986. Virus, Mycoplasma and Rickettsia Diseases of Fruit Trees. Martinus Nijhoff, Dordrecht, The Netherlands. 841 pp.

Nyland, G., Gilmer, R. M., and Moore, J. D. 1976. Prunus ringspot group. Pages 104-132 in: Virus Diseases and Noninfectious Disorders of Stone Fruits in North America. U.S. Dep. Agric. Agric. Handb. 437.

(Prepared by G. I. Mink)

Viruses or Pathogens Spread by Mites, Insects, and Nematodes

Cherry Mottle Leaf Virus

Cherry mottle leaf disease was first reported in Oregon in 1920. Since then, mottle leaf has been observed in California, Idaho, Montana, Oregon, Utah, Washington, and British Columbia. A sweet cherry disease called mottle leaf reported from the former Yugoslavia is not related to North American cherry mottle leaf. The disease can be very severe locally.

Symptoms

Characteristic symptoms on most cherry cultivars consist of light green or whitish, diffuse discoloration between the secondary leaf veins. The transition from chlorotic to healthy green tissue tends to be gradual, in contrast to symptoms induced by Prunus necrotic ringspot virus, plum dwarf virus, and other viruses. Affected leaves often develop an irregular shape. During the early part of the growing season, the outline of leaves can be severely affected, to the extent that the margin may appear shredded (Plate 113). However, symptoms tend to become less severe on later-developing leaves. There is a wide range of strains of different severity. Some reduce fruiting, while others reduce fruit size and delay ripening. Long-established infections can lead to severe stunting and crop failure. Frost peach is a symptomless host.

Causal Agent

The disease agent was known previously to be vectored by the bud mite *Eriophyes inaequalis* Wilson & Oldfield. More recently, a closterovirus with a modal length of 760 nm, a width of 10.2 nm, and striations 3.1 nm in pitch was sap-transmitted from diseased cherry to *Chenopodium quinoa*. Virus particles contained single-stranded RNA with a molecular weight of 2.79 $\times 10^6$. Purified virus preparations caused mottle leaf symptoms following back-inoculations to Bing sweet cherry. Although pathogenicity tests were done, experimentation demonstrating the relationship between the closterovirus isolate and the mottle leaf disease vector is lacking.

Disease Cycle and Epidemiology

The mottle leaf agent occurs naturally in bitter cherry in the northwestern United States and British Columbia. From this reservoir host, the virus is carried by the vector bud mite to sweet cherry, peach, and apricot. The disease agent also has been found in plum, almond, mahaleb, Chinese wild peach, and Nanking cherry. Symptom development is most pronounced on sweet cherry, especially during spring and early summer. Grafting or budding healthy indicators, such as Bing sweet cherry, with mottle leaf-infected material will result in typical symptoms within one year or less.

Control

Virus-free material can be produced readily from infected sources by heat treatment. Use of virus-free certified understock and budwood is an excellent preventive measure. In areas where field transmission occurs, bitter cherry and infected orchard trees should be removed.

Selected References

Babovic, M. V., Perisic, M., and Matijevic, M. 1981. Cherry mottle leaf. Acta Hortic. 94:129-133.

Hansen, A. J. 1978. Natural occurrence of cherry mottle leaf in apricot. Plant Dis. Rep. 62:630-632.

Howell, W. E., and Norton, R. A. 1993. 'Frost' peach: A symptomless source of cherry mottle leaf disease. Fruit Var. J. 47(2):125-126.

James, D., and Mukerji, S. 1993. Mechanical transmission, identification, and characterization of a virus associated with mottle leaf in cherry. Plant Dis. 77:271-275.

Oldfield, G. N. 1970. Mite transmissions of plant viruses. Annu. Rev. Entomol. 15:343-380.

Reeves, E. L. 1941. Mottle leaf, a virus disease of cherries. J. Agric. Res. 62:555-572.

(Prepared by A. J. Hansen and G. N. Oldfield)

Peach Mosaic

Peach mosaic was first observed in 1931 in Texas and Colorado. Later surveys documented its occurrence in Arizona, Arkansas, southern California, New Mexico, Oklahoma, and Utah, and in several states of Mexico. Peach mosaic has been confused with peach calico, plum pox, and other plum virus diseases in Europe. Desvignes noted strong similarities between peach mosaic and peach latent mosaic in Europe and peach yellow mosaic in Japan; however, the relationships among these three diseases still remain to be determined. In Spain, Flores and Llacer reported an association of a viroid with peach latent mosaic. Recently, a closterovirus-like particle was sap-transmitted from a peach mosaic–infected tree to herbaceous hosts, and it reacted with monoclonal antibodies produced to cherry mottle leaf virus.

The highly infectious nature of peach mosaic required implementation of stringent control programs in the United States in 1934. During the first five years of these programs, more than 204,000 infected peach trees in California and 58,000 in Colorado were destroyed. By 1955, some 125,000 infected peach trees had been destroyed in Mesa County, Colorado. These eradication efforts successfully reduced peach mosaic to very low levels by the early 1960s, and only a few confirmed sightings were reported during the 1980s. No parallel effort was implemented in Mexico, and the incidence in Chihuahua remains extremely high. More recently, the mite that vectors the disease was found in other peach-growing areas of Mexico.

Symptoms

Disease symptoms vary with the host species, the virus isolate, and coinfections with other stone fruit viruses. On diseased trees, color-breaking of flower petals, seen primarily in peach cultivars with large, showy, pink flowers, such as Carman, Fay Elberta, Loring, and Rio Oso Gem, consists of radially oriented pale pink to white islets and/or streaks that develop along the petal base toward the outer edge (Plate 114). Symptomatic petals are smaller, crinkled, and deformed.

Budbreak is retarded on infected trees. Leaves show chlorotic spots that vary from tiny flecks to streaks to vein feathering or bold blotches (Plate 115). These symptoms are best viewed with the aid of backlight illumination of fully expanded leaves. The foliage of infected trees turns a darker green by midsummer, when any of the severely symptomatic leaves may be small, narrow, and asymmetric, with deformed, crinkled margins. The chlorotic areas become necrotic and drop out, producing a shot-hole and tattered appearance. Twigs and shoots are dwarfed and appear as a stubby growth. On severely affected branches, the linear growth can be rosetted, and shoot proliferation may occur toward the shoot apex, presenting a witches'-broom appearance. Fruit of freestone peaches and nectarines are commonly deformed and unmarketable. The fruit surface along the suture side becomes rough before the stone-hardening stage (Plate 116).

Cultivated European and Japanese plums are highly susceptible to the disease agent and express only the leaf symptoms described for peach. Susceptible wild plums, such as American plum, Mexican plum, and wild-goose plum, rarely develop symptoms. In apricot, the more sensitive cultivars like Blenheim develop leaf discoloration patterns and distortion and stubby twig symptoms. They also tend to be less vigorous, produce less fruit, and have more sunburned fruit than healthy trees. Almond trees generally exhibit mild foliar and fruit symptoms, with appreciable reduction in yield or nut quality.

Causal Agent

The causal agent of peach mosaic is easily graft-transmitted to healthy peach using fruit, leaf, root, or bud tissues and a contact period as short as two days. Attempts to transmit the agent through dodder have failed. At least three strains of the agent have been recognized. When these were graft-inoculated into *Prunus,* a severe peach strain also caused symptoms in plum but not in almond or apricot. A plum strain occurring naturally in wild plum in the southwestern United States induced symptoms in all three hosts. A mild strain incited faint or no symptoms in any of the hosts. The mild strain provided cross-protection against the severe peach strain but not against the plum strain.

The vector is the peach bud mite, *Eriophyes insidiosus* Keifer & Wilson, which feeds and reproduces on developing leaf primordia within the bud. A single infectious mite can transmit the disease agent to a healthy tree. The mite is windborne.

Disease Cycle and Epidemiology

Natural spread of peach mosaic via the vector mite has occurred from peach to peach, apricot, and almond and from wild plum to peach. Native plums such as American plum, Mexican plum, and wild-goose plum are symptomless hosts. Sweet cherry is immune.

Control

Previously, a quarantine against importation of disease-tolerant peach or nectarine cultivars was imposed because these could become infected, remain symptomless, and serve as reservoirs. Peach mosaic is best controlled through planting virus-certified nursery stock and by prompt detection and removal of diseased trees. Also, the mite vector has been controlled with prebloom and petal-fall miticide applications. However, this approach requires caution to minimize the risks to honeybees, which are generally active during the spring season.

Selected References

Cochran, L. C., and Pine, T. S. 1958. Present status of information on host range and host reactions to peach mosaic virus. Plant Dis.

Rep. 42:1225-1228.
Desvignes, J. C. 1986. Peach latent mosaic and its relation to peach mosaic and peach yellow mosaic virus diseases. Acta Hortic. 193:51-57.
Flores, R., and Llacer, G. 1988. Isolation of a viroid-like RNA associated with peach latent mosaic disease. Acta Hortic. 235:325-332.
James, D., and Howell, W. E. 1993. Comparisons of cherry mottle leaf and a virus associated with peach mosaic disease. Abstr. 17.1.20. Int. Congr. Plant Pathol. 6th, Montreal, Canada, p. 304.
Jones, L. S., Wilson, N. S., Burr, W., and Barnes, M. M. 1970. Restriction of the peach mosaic virus spread through control of the vector mite, *Eriophyes insidiosus*. J. Econ. Entomol. 63:1551-1552.
Oldfield, G. N. 1970. Mite transmission of plant viruses. Annu. Rev. Entomol. 15:343-380.
Oldfield, G. N., Creamer, R., Gispert, C., Osorio, F., Rodriguez, R., and Perring, T. M. 1995. Incidence and distribution of peach mosaic and its vector, *Eriophyes insidiosus* (Acari: Eriophyidae) in Mexico. Plant Dis. 79:186-189.
Pine, T. S. 1965. Host range and strains of peach mosaic virus. Phytopathology 55:1151-1153.
Pine, T. S. 1976. Peach mosaic. Pages 61-70 in: Virus Diseases and Noninfectious Disorders of Stone Fruits in North America. U.S. Dep. Agric. Agric. Handb. 437.
Richards, B. L., and Cochran, L. S. 1957. Virus and virus-like diseases of stone fruits in Utah. Utah Agric. Exp. Stn. Bull. 384. 130 pp.

(Prepared by H. J. Larsen and G. N. Oldfield)

Little Cherry

Little cherry received major attention in British Columbia, Canada, after the causal agent was introduced in 1933 with infected Japanese flowering cherry (*Prunus serrulata* 'Shirofugen'). Circumstantial evidence indicates that the disease originated in Japan, where the agent is common in many flowering cherry cultivars. Little cherry has been reported from many cherry-growing countries in Europe as well as from Japan (where it is likely endemic), Canada, and the United States. Currently, there is no explanation for the relatively high incidence of the disease in England versus the lack of incidence in Germany.

Symptoms

Fruits of sweet cherry are more severely affected than any other part of the tree. In the first year after infection, fruit is about half normal size (Plate 117) and triangularly shaped when viewed from the bottom. The fruit skin on two of the three sides tends to shrivel slightly, causing a leathery appearance. Fruit color develops normally until approximately 10 days before harvest, when normally dark-colored cultivars do not develop beyond the bright red stage. Flavor is bland but not bitter, which is one way to distinguish between little cherry and the mycoplasma-induced X-disease of cherry.

Host reaction among cultivars of *P. avium* is variable. Fruits of Lambert and Black Republican are most severely affected; those of Bing, Black Tartarian, mazzard F12/1, and Montmorency (a sour cherry) are moderately affected, and symptoms on Van, Deacon, and Royal Ann-type sweet cherries are mild. Foliage tends to be light in color, and tree growth is often reduced. In subsequent years, symptoms are less severe, and infected trees may be hard to identify in the field.

Causal Agent

Flexuous, rod-shaped particles have been associated consistently with little cherry–infected trees and are thought to be the causal agent of the disease. The particles consist of a characteristically open, electron-transparent ultrahelix with a pitch of 4.25 nm and a prominent, opaque central core. Particle width is 10.6 nm, and the predominant length is 1.5 μm. Once purified, the particles are very stable. Particle levels in infected trees are very low.

Disease Cycle and Epidemiology

The only proven vector of little cherry is the apple mealybug, *Phenacoccus aceris* Sign., which is widespread in Europe and has become established in many parts of North America. Field spread of the disease is closely related to the population density of the vector. Since it is very slow-moving, new infections tend to occur in trees adjoining older infected ones. The agent affects all species and cultivars in the *P. serrulata* group, which includes sweet and sour cherry, Japanese flowering cherry, bitter cherry, mahaleb, and pin cherry. Commercial cultivars of peach, plum, and almond and their ornamental relatives are immune to little cherry, as are Nanking cherry and chokecherry.

Detection

Field testing requires inducing visual symptoms on fruiting trees or in foliage of young indicator trees. In British Columbia, healthy trees budded in August with infected buds display prominent fruit symptoms the next summer, especially in severely affected indicators such as the sweet cherry cultivar Lambert. Similar budding of one- or two-year-old foliage indicator trees induces prematurely dark red coloration approximately six weeks ahead of normal leaf fall. The leaves usually retain green or yellow coloration along the veins, and this characteristic can be used to distinguish little cherry from similar coloration induced by certain strains of prune dwarf virus or other causes. The specifically developed indicator cultivar Canindex 1, as well as Deacon and Sam (all *P. avium*) (Plate 118), is being used in routine detection programs for little cherry.

For microscopic detection, wall deposits in disease-related phloem cells of infected trees can be stained with acridine orange and observed with the fluorescent microscope. Flexuous rods about 12 nm in diameter and small vesicles associated with diseased cells, about 75 nm in diameter, frequently have been observed in phloem parenchyma and companion cells of leaf midribs and petals. These structures can be visualized by electron microscopy in ultrathin sections stained with uranyl-acetate and lead citrate. Neither the fluorescent microscope nor electron microscopy is used routinely for detection of the causal agent.

Control

Because the natural hosts of little cherry are limited and the vector moves slowly, control of the disease is relatively easy. A combination of vector control by insecticide application or by release of predators *and* eradication of known infected trees controlled disease spread in the area of a major outbreak, namely, the Kootenay Valley of British Columbia. Continued efforts will likely lead to eradication of the disease. However, attention must be paid to potential natural hosts such as bitter cherry and pin cherry, which could become infected by vector transmission.

Prohibition of the importation of little cherry–infected ornamental cherries and the use of virus-free understock and budwood have greatly reduced the spread of little cherry in British Columbia in recent years.

Selected References

Hansen, A. J. 1985. Canindex 1, a superior indicator cultivar for little cherry disease. Plant Dis. 69:11-12.
Legrande, G., and Verhoyen, M. 1985. Détection du little cherry disease chez le cerisier par des observations au microscope à lumière u.v. Verh. Fac. Landbouwwet. Rijksuniv. Gent. 50/3b:1283-1294.
Ragetli, H. W. J., Elder, M., and Schroeder, B. K. 1982. Isolation and properties of filamentous virus-like particles associated with little cherry disease in *Prunus*. Can. J. Bot. 60:1235-1248.
Raine, J., McMullen, R. D., and Forbes, A. R. 1986. Transmission of the agent causing little cherry disease by the apple mealybug *Phenacoccus aceris* and the dodder, *Cuscuta lupuliformis*. Can. J. Plant Pathol. 8:6-11.
Raine, J., Weintraub, M., and Schroeder, B. 1975. Flexuous rods and vesicles in leaf and petiole phloem of little cherry diseased *Prunus*

spp. Phytopathology 65:1181-1186.
Verbeek, R. B. 1977. The occurrence of wall deposits in phloem cells of cherry trees infected with little cherry disease. Can. J. Bot. 55:2344-2347.
Wilks, J. M., and Welsh, M. F. 1955. Sweet cherry foliage indicator hosts for the virus that causes little cherry. Can. J. Agric. Sci. 35:595-600.

(Prepared by A. J. Hansen)

Plum Pox Virus

Plum pox virus (also known as sharka) was first seen in southwest Bulgaria about 1918. Since then, the disease has been widely disseminated in planting material to most of Europe with the exception of Scandinavia. It causes serious losses in plum (*Prunus domestica*), peach, nectarine, and apricot.

Symptoms

Foliar symptoms in plum consist of pale green chlorotic spots, rings, and lines that are visible from early summer onward (Plate 119). These markings can become necrotic. Symptoms frequently are restricted to only a few leaves per shoot. Infected trees usually are not stunted and are difficult to identify.

Fruit symptoms in red and dark-colored plums consist of rings and blotches in fully expanded unripe fruit (Plate 120). These markings tend to disappear as the fruits color and reach maturity. Sunken rings and spots occur in some cultivars, overlying discolored flesh. Red rings and spots can occur on the stones. Affected fruit are low in sugar and tasteless; they drop from the tree prematurely. An abundance of abscised fruit helps in locating suspect trees in infected orchards.

In peach, symptoms appear in the first leaves to expand and consist of chlorotic veinclearing and veinbanding, with twisting and distortion of the lamina. Symptoms in peach fruit (pale or dark rings, lines, and spots) may disappear at ripening. Affected fruits sometimes are deformed.

In apricot, leaf symptoms are similar to those in plum but are less conspicuous. Fruit of some cultivars, however, are severely marked and may be misshapen, with flavorless, dry flesh around the stone, which is usually marked with pale rings, lines, or spots (Plate 121).

In general, cultivars of each host show differential responses to infection.

Causal Agent

Plum pox virus is a member of the potyvirus (potato virus Y) group, characterized by its flexuous, filamentous particles (approximately 750 nm long) that contain a single-stranded RNA with a molecular weight of 3.5×10^6. The virus is aphid-transmitted in a nonpersistent manner, and infection is associated with pinwheel inclusion bodies in the cytoplasm.

Disease Cycle and Epidemiology

The most important means of spread is by diseased plant material, which accounts for much of the rapid spread among and within European countries during the last 20 years. Once established, plum pox virus is extremely difficult to eradicate. Spread by aphids usually occurs along a comparatively steep gradient; i.e., new infections next to infected sources and occasionally at longer distances. Infection is slow to become systemic and is often confined to one or two limbs of a tree. Because symptoms can be mild or transient, considerable spread can occur from new disease foci before infections are noticed. There is no detailed information on the epidemiology of plum pox virus (e.g., the relative importance of different vector aphids, seasonal variation in spread, or the effect of pesticides).

Within the genus *Prunus,* the virus infects most species in the subgenera *Prunophora* (plums and apricots) and *Amygdalus* (almonds and peaches). Some members of the *microcerasus* section of the subgenus *Cerasus* can be infected experimentally, but not members of the *eucerasus* (cherries), *pseudocerasus* (ornamentals), *padus* (bird cherries), or *laurocerasus* (cherry laurel) sections. There are, however, reports of the virus occurring naturally in sweet and sour cherries in Moldavia and Bulgaria. Few species other than *Prunus* species have been naturally infected, but the virus has a wide experimental host range, including *Chenopodium foetidum* Schrad., *Lamium amplexicaule* L., *Nicotiana clevelandii* Gray, *N. megalosiphon* Heurck & Muell., *Pisum sativum* L., *Ranunculus arvensis* L., *Senecio sylvaticus* L., and *Zinnia elegans* Jacq.

Control

Control depends on obtaining healthy plant material, achieved in countries in Europe by certification programs. Trees are grown from tested propagation material in isolation from potential sources of disease and are inspected before distribution to growers. The most widely used test for plum pox virus is enzyme-linked immunosorbent assay (ELISA), but results should be interpreted cautiously because distribution of virus is unpredictable and uneven in infected trees. Besides ELISA, molecular-based assays such as polymerase chain reaction are being done in many countries for plum pox virus detection and identification.

Alternatively, standard biological indicators of plum pox virus have included seedlings of either peach cultivars GF 305 and Siberian C or Nanking cherry (*P. tomentosa*). Recently, eight plum pox isolates were compared in controlled inoculations under standard growing conditions in a quarantine containment facility in Frederick, Maryland. In Nanking cherry, leaf symptoms—consisting of chlorotic veinclearing (Plate 122), chlorotic oak-leaf patterns, and mild distortion among scattered leaves—developed in 30 days. Nanking cherry was found to be superior to the peach indicators in detecting all virus isolates tested.

The productivity of orchards with low levels of infection can be prolonged by frequent inspection and removal of infected trees. Rigorous spraying for aphids may delay the spread of virus. In areas where plum pox virus is common, the use of tolerant cultivars is the only effective control. Although a few immune or highly resistant plum and apricot cultivars exist, they are not in use commercially.

Selected References

Adams, A. N. 1978. The detection of plum pox virus in *Prunus* species by ELISA. Ann. Appl. Biol. 90:215-221.
Maiss, E., Timpe, U., Brisske, A., Jelkmann, W., Casper, R., Himmler, G., Mattanovich, D., and Katinger, H. W. D. 1989. The complete nucleotide sequence of plum pox virus RNA. J. Gen. Virol. 70:513-524.
Mathys, G., ed. 1974. Progrès réalisés dans la connaissance de la sharka. OEPP Bull. 4. 125 pp.
Van Oosten, H. J. 1975. Susceptibility of some woody plant species, mainly *Prunus* spp., to sharka (plum pox) virus. Neth. J. Plant Pathol. 81:199-203.
Varveri, C., Ravelonandro, M., and Dunez, J. 1987. Construction and use of a cloned cDNA probe for the detection of plum pox virus in plants. Phytopathology 77:1221-1224.

(Prepared by A. N. Adams; information on the Nanking cherry test for plum pox virus was supplied by V. Damsteegt)

Cherry Rasp Leaf Virus

Cherry rasp leaf disease was first found in Colorado in 1935 and now occurs in most cherry-growing districts west of the Rocky Mountains of North America. The virus is spread by the dagger nematode *Xiphinema americanum* Cobb and occurs typically in patches of a few infected trees in otherwise healthy orchards. Circumstantial evidence indicates that the causal virus is present in native host plants. New outbreaks typically occur on land that has recently been cultivated.

Symptoms

In cherry, leaflike projections (enations) develop along the midrib on the underside of affected leaves (Plate 123). In severe cases, the whole leaf can be affected. All cherry cultivars seem to react similarly. Disease symptoms are initially restricted to the lower branches and limbs of mature trees because the virus is introduced into the roots. Severely infected branches tend to be more winter-tender than the rest of the tree. Eventually, the diseased tree declines. Peach trees infected with cherry rasp leaf virus (CRLV) also decline and die.

Causal Agent

CRLV is a spherical virus (28 nm in diameter) with a bipartite genome of single-stranded RNAs with molecular weights of 2.0 and 1.5×10^6. The protein shell is composed of two polypeptide molecules with molecular weights of 22,500 and 24,000. The virus is a member of the nepovirus group and is vectored by the dagger nematode; it is borne in seeds of *Chenopodium quinoa* and dandelion (*Taraxacum officinale*), which produce infected seedlings; it is easily sap-transmitted, and it is infectious across a wide range of herbaceous and woody plants. CRLV is moderately immunogenic and is unrelated serologically to recognized members of the nepovirus group, e.g., tobacco ringspot, tomato ringspot, cherry leaf roll, peach rosette mosaic, raspberry ringspot, tomato black ring, and arabis mosaic viruses.

Disease Cycle and Epidemiology

CRLV is easily graft-transmitted to healthy cherry trees. In orchards, however, it is spread by the dagger nematode. The common cherry rootstocks mahaleb (*Prunus mahaleb*) and mazzard (*P. avium*) are susceptible to CRLV. CRLV-infected cherry, peach, and apple trees have been found in commercial orchards (CRLV causes leaf enation and flat apple diseases in apple). Natural latent infections also occur in a number of weed species common in diseased orchards in western North America, including balsam root (*Balsamorhiza sagittata* (Pursh) Nutt.), dandelion, plantain (*Plantago major* L.), and bindweed (*Convolvulus arvensis* L.). The virus was transmitted to 10–20% of seeds of *C. quinoa* and dandelion.

Cherry replants in dagger nematode-infested soils may become infected. Dagger nematodes recovered from infested field soils transmitted CRLV to bait plants of cucumber and mazzard cherry.

Control

Although meticulous experiments have not been reported, use of fumigants for nematode vector control and elimination of known weed hosts of CRLV from the orchard might be effective in controlling cherry rasp leaf disease.

Selected References

Hansen, A. J., Nyland, G., McElroy, F., and Stace Smith, R. 1974. Origin, cause, host range, and spread of cherry raspleaf disease in North America. Phytopathology 64:721-727.

Nyland, G. 1976. Cherry raspleaf. Pages 219-221 in: Virus Diseases and Noninfectious Disorders of Stone Fruits in North America. U.S. Dep. Agric. Agric. Handb. 437.

Parish, C. L. 1977. A relationship between flat apple disease and cherry raspleaf disease. Phytopathology 67:982-984.

Stace Smith, R., and Hansen, A. J. 1976. Cherry raspleaf virus. Descriptions of Plant Viruses, No. 159. Commonwealth Mycological Institute and Association of Applied Biologists, Kew, Surrey, England. 4 pp.

(Prepared by A. J. Hansen)

Peach Rosette Mosaic Virus

Rosette mosaic disease of peach was noticed first in Michigan peach orchards in 1917 and was described as of virus etiology in 1933. It is also present in Ontario, Canada. Peach rosette mosaic virus (PRMV) also causes a disease in grape (*Vitis labrusca* L. 'Concord').

Symptoms

Leaf emergence of infected peach trees is delayed, then leaves develop chlorotic mottling. Leaf distortion may follow a chlorotic lesion that forms on one side of the leaf lamina. Leaves formed later in the season are narrower than healthy leaves. A shortening of internodes (rosette) is a prominent symptom (Plate 124) in late spring and summer. Infected trees appear darker green, are stunted, and produce little or no fruit. In the field, a latent period of nine to 21 months passes before symptoms appear.

Causal Agent

PRMV is a multicomponent, polyhedral virus measuring 28 nm in diameter, with two species of single-stranded RNAs with molecular weights of 2.2 and 2.4×10^6 and a single protein species of molecular weight 57,000. PRMV is a member of the nepovirus group and is vectored by the nematodes *Xiphinema americanum* Cobb and *Longidorus diadecturus* Eveleigh & Allen. The virus is moderately immunogenic and forms a single precipitin band in gel diffusion serological tests. PRMV is antigenically distinct from other nepoviruses.

Disease Cycle and Epidemiology

PRMV can be transmitted by budding, but natural spread in peach orchards and vineyards is by nematode vectors. Other natural hosts include dandelion (*Taraxacum officinale*), Carolina horsenettle (*Solanum carolinense* L.), and curly dock (*Rumex crispus* L.). PRMV is mechanically transmissible from young leaves of peach or grapes (tissues extracted in phosphate-nicotine buffer) to the indicator host *Chenopodium quinoa*. *C. amaranticolor* also is a useful assay plant. The virus is seed-borne in *C. quinoa*.

Control

Roguing of diseased trees and preplant fumigation are recommended. High rates of the fumigant injected both shallow (15–20 cm) and deep (75–90 cm) during the fall season may be required to effectively reduce vector populations in infested soils. Rootstocks at planting should be certified as virus-free.

Selected References

Allen, W. R., Van Schagen, J. G., and Eveleigh, E. S. 1982. Transmission of peach rosette mosaic virus to peach, grape, and cucumber by *Longidorus diadecturus* obtained from diseased orchards in Ontario. Can. J. Plant Pathol. 4:16-18.

Cation, D. 1933. An infectious rosette on peach trees. Mich. Agric. Exp. Stn. Q. Bull. 16:79-84.

Dias, H. F. 1975. Peach rosette mosaic virus. Descriptions of Plant Viruses, No. 150. Commonwealth Mycological Institute and Association of Applied Biologists, Kew, Surrey, England. 4 pp.

Dias, H. F., and Cation, D. 1976. The characterization of a virus responsible for peach rosette mosaic and grape decline in Michigan. Can. J. Bot. 54:1228-1239.

Ramsdell, D. C., and Myers, R. L. 1978. Epidemiology of peach rosette mosaic virus in a Concord grape vineyard. Phytopathology 68:447-450.

(Prepared by D. C. Ramsdell)

Tomato Ringspot Virus

Tomato ringspot virus (TmRSV) causes or is closely associated with diseases that occur in a range of *Prunus* species, including peaches, plums, and cherries. Diseases caused by TmRSV are given various names that often designate the host species that is affected and the symptoms induced. Examples include Prunus (peach) stem pitting, peach yellow bud mosaic, cherry eola rasp leaf, prune brown line, and constriction and decline of Stanley prune. The virus is endemic in North America and is found in *Prunus* spp. throughout the northeastern and Middle Atlantic states, in California, and to a lesser extent elsewhere in the United States.

Symptoms

Disease symptoms vary among *Prunus* spp. but are generally characterized by two conditions: first, affected areas show various phloem disorders near the ground line that result in thickened bark and pitting of the young xylem under the bark; and second, necrotic phloem tissue develops at the graft union in trees that have TmRSV-resistant scions growing on virus-susceptible rootstocks. Severity of symptoms varies according to the length of time that plants have been infected, the cultivar, host species, and virus isolate.

Prunus stem pitting—In peach orchards of the eastern United States, trees affected by Prunus stem pitting disease show an overall unthrifty appearance. The leaves develop a droopy posture, turn prematurely yellow or red, and abscise prematurely. In spring, unfolding of leaf buds is delayed. All of these symptoms can be confused with other conditions, such as root rot. However, the most reliable field symptoms of stem-pitted trees are abnormally thickened, spongy bark and pits and grooves in the wood beneath (Plate 125). These symptoms are often more severe in the trunk area immediately above and below the soil line. Symptom severity varies according to the cultivar. For example, peach cultivars Red Haven and Redskin develop deep grooves, whereas Elberta peach forms short, shallow pits. For certain cultivars, symptomatic trees that are severely stunted begin to decline and later die; however, some infected trees may remain alive for several years.

In sweet and sour cherries, the symptoms of Prunus stem pitting disease are similar to those described for peach trees, and severity depends on the scion-rootstock combination. Infected mahaleb (*P. mahaleb*) roots develop numerous short and relatively shallow pits, while those of Stockton Morello (*P. cerasus*) develop deep, extensive grooves and cambial necrosis. Sweet cherry cultivars such as Bing also exhibit severe pits, but the virus remains latent in cultivar Royal Ann (Napoleon). On infected Royal Ann, the stem-pitting symptom is found only in the rootstock portion of the tree.

Peach yellow bud mosaic—In peach orchards of California, symptoms of peach yellow bud mosaic contrast with those just described. Growth of infected leaf buds is severely limited, and leaf clusters are small and yellow (Plate 126). Symptomatic leaves die as seasonal temperatures rise. During the first year of infection, the leaves have chlorotic spots, rings (Plate 127), or oak-leaf patterns (mosaic phase). In the second year, the shoots exhibit yellowed buds (yellow bud phase) that generally die during the fruiting season. Affected limbs have fewer normal leaves, and fruit production is reduced. Similar symptoms have been observed in sweet cherry and almond trees infected with the yellow bud mosaic strain of TmRSV. Trunks of diseased trees develop, at best, scattered pockets of shallow pits.

Brown line—Prune trees infected with TmRSV grow poorly, with sparse foliage. When Stanley prune is grafted onto Myrobalan (*P. cerasifera* Ehrh.) rootstock, the area just below the graft union grows more slowly than the scion portion of the tree, producing an overgrowth appearance. Removal of the bark reveals necrotic tissues (a brown line) at the scion-rootstock junction (Plate 128). Trees die following onset of brown line symptoms, indicative of a hypersensitive response. In California, brown line disease is observed in French prune and Italian prune on Myrobalan and peach roots.

Yellow bud mosaic—Fruit-marking symptoms usually are not associated with TmRSV. An exception is almond, on which the nut hulls become thickened and wrinkled when infected by the yellow bud mosaic strain of TmRSV.

Causal Agent

TmRSV has a bipartite genome with RNA molecular weights of 2.17 and 2.05×10^6. Virus particles are spherical, about 28 nm in diameter, with a protein shell composed of single polypeptide units with a molecular weight of 58,000. TmRSV is a member of the nepovirus group and is transmitted by the dagger nematodes *Xiphinema americanum* Cobb, *X. rivesi* Dalmasso, and *X. californicum* Lamberti & Bleve-Zacheo. Both virus and vectors are endemic in the northeastern United States. TmRSV antisera do not cross-react with other members of the nepovirus group.

Disease Cycle and Epidemiology

TmRSV is transmitted by grafting and budding. Natural spread in orchards and nurseries results directly from transmission by nematode vectors. In infested sites, the nematode and virus can persist for years on roots of infected perennial and herbaceous plants. In addition to *Prunus* species, other woody hosts, e.g., grapevine (*Vitis* spp.) and raspberry (*Rubus* spp.), are hosts of TmRSV. Nanking cherry (*P. tomentosa*) is a sensitive indicator for TmRSV. The virus also infects a wide range of common orchard weeds such as dandelion, sheep sorrel (*Rumex acetosella* L.), and common chickweed (*Stellaria* spp.). The virus is seedborne in several weed hosts.

Control

Before nursery certification programs were established, nursery stock was a common source of infected trees. However, under stringent protocols adopted by the nursery industry and state agencies, production of virus-tested trees has greatly reduced the problem. Hence, growers should obtain and plant trees derived from a virus certification program.

Thorough land preparation also is important because weeds can serve as a virus reservoir. Weeds should be destroyed and prevented from becoming reestablished. If nematode vectors are detected, the orchard site should be fumigated.

Perhaps the most economical method for controlling TmRSV is use of a resistant rootstock. In California, Marianna 2624 plum (*P. cerasifera* × *P. munsoniana*), a clonally propagated rootstock, is resistant to TmRSV and is used routinely when new plantings follow diseased prune orchards. Stanley prune also is highly resistant and could be a useful rootstock for plum and prune cultivars. Resistance among rootstocks commonly used for cherry and peach is unknown.

Selected References

Cummins, J. N., and Gonsalves, D. 1986. Constriction and decline of prune associated with tomato ringspot virus. J. Am. Soc. Hortic. Sci. 11:315-318.

Forer, L. B., Powell, C. A., and Stouffer, R. F. 1984. Transmission of tomato ringspot virus to apple rootstock cuttings and to cherry and peach seedlings by *Xiphinema rivesi*. Plant Dis. 68:1052-1054.

Hoy, J. W., and Mircetich, S. M. 1984. Prune brown line disease: Susceptibility of prune rootstocks and tomato ringspot virus detection. Phytopathology 74:272-276.

Mircetich, S. M., and Fogle, H. W. 1976. Peach stem pitting. Pages 77-87 in: Virus Diseases and Noninfectious Disorders of Stone Fruits in North America. U.S. Dep. Agric. Agric. Handb. 437.

Powell, C. A., Forer, L. B., Stouffer, R. F., Cummins, J. N., Gonsalves, D., Rosenberger, D. A., Hoffman, J., and Lister, R. M. 1984. Orchard weeds as hosts of tomato ringspot and tobacco ringspot viruses. Plant Dis. 68:242-244.

Schlocker, A., and Traylor, J. A. 1976. Yellow bud mosaic. Pages 156-165 in: Virus Diseases and Noninfectious Disorders of Stone Fruits in North America. U.S. Dep. Agric. Agric. Handb. 437.

Stouffer, R. F., Soulen, D. M., and Smith, S. H. 1975. Spread and control of Prunus stem pitting. Acta Hortic. 44:107-112.

(Prepared by D. Gonsalves)

Other Nepoviruses

Economically important nepoviruses of stone fruits in Europe include arabis mosaic virus (ArMV), cherry leaf roll virus (CLRV), raspberry ringspot virus (RRSV), strawberry latent ringspot virus (SLRV), and tomato black ring virus (TBRV). Each virus alone may cause obvious damage, but usually the diseases are more significant when these nepoviruses occur in combination with other common stone fruit viruses such as Prunus necrotic ringspot virus (PNRSV) or prune dwarf virus (PDV). In the future, these nepoviruses may have a significant impact in areas outside Europe. SLRV already infects cherry in Canada and New Zealand, and CLRV and RRSV infect cherry in Turkey.

Symptoms

Apricot—In the former Czechoslovakia, a disease characterized by bare twigs and unfruitfulness has caused heavy losses in apricot orchards. Lateral shoots and spurs die and drop off, leaving a rosette of twigs at the tip of each branch. Leaves are few and often roll upward toward the midrib; petioles may be turned to expose the undersides of affected leaves. In advanced stages of the disease, flowering and fruit set are reduced, flowers and fruit abscise prematurely, and the remaining fruit may be deformed. These symptoms are associated with the presence of SLRV and cucumber green mottle mosaic virus.

Cherry—CLRV causes chlorotic leaf spots, leaf curling, shoot dieback, bark splitting, gumming, and decline of sweet and sour cherries. In sweet cherry, bloom and leaf emergence may be delayed and weakened. Leaves are rolled, rigid, and sometimes discolored. Enations develop on the undersides of leaves in trees infected only with CLRV but are more numerous if PDV is also present. Both sweet and sour cherries infected with CLRV alone may die within five years; decline is more rapid if the tree is also infected with PNRSV.

The European cherry rasp leaf disease has caused severe tree losses in many European countries. In most cases, sweet cherry trees with rasp leaf symptoms were infected with ArMV, CLRV, RRSV, or SLRV, or any of these viruses in combination with PDV. In addition to leaf enations, the leaves develop chlorotic areas, wavy and distorted leaf blades with deep serrations, and cleared, protruding veins. Leaf emergence may be delayed, but abscission may occur prematurely. Shoot growth is reduced to the extent that tips of many shoots appear as rosettes of buds and leaves. Fruit development is usually normal but prolonged; occasionally, a fruit develops unevenly, with deep sutures and pointed tips. Yield is usually decreased. Cankers, gumming, and then dieback occur in twigs and branches; infected trees eventually die.

Peach—In Germany, peach trees infected with TBRV have yellowish wavy and distorted leaves with irregular chlorotic spots, stunted shoot growth with shortened internodes, and rosettes of leaves at the tips of stunted shoots. Leaf symptoms become masked during summer, but fruit production is reduced. Dieback of affected branches precedes tree death.

SLRV has been isolated from peach trees with willow leaf rosette in Italy, *court noue* in France, and shoot dwarfing in Germany. These diseases are most obvious in spring and may be mild later. Symptoms include chlorotic leaf spots, small narrow leaves, and rosettes of thick, narrow, dark green leaves with yellowish veins. Flowering and leaf emergence are delayed, and leaf abscission on rosetted shoots may be premature. Shoot growth may be reduced because of short internodes at the shoot base. Eventually, branches die back. Fewer fruit, which may be small and bumpy, are produced on infected trees.

ArMV has been detected in peaches during testing of seed source trees in Hungary, but no adverse economic impact of this virus on peach trees has been reported.

Plum—In Poland, RRSV causes mild symptoms in plum, but the combination of plum pox virus and RRSV results in severe damage in certain cultivars. Young shoots become flattened and develop longitudinal cracks. Necrosis develops at the graft union, and the lower parts of the trunk become thickened. Infected trees are stunted, bear sparse foliage, and eventually decline. Myrobalan latent ringspot virus (MLRV) has been isolated from Myrobalan B plum rootstocks in France. Symptoms include retarded foliation and reduced shoot growth. The economic impact of this nepovirus is unknown.

Causal Agents

Nepoviruses share many common properties, including similar polyhedral particle morphologies of 28-nm diameter. Viral particles separate in gradient columns as two or three components, designated top, middle, and bottom. All components have the same protein subunits, with molecular weights of $5.5–6.0 \times 10^3$. The genome consists of two positive-sense, single-stranded molecules of RNA with molecular weights of $1.3–2.4 \times 10^6$ and 2.6×10^6; both are required for infection. They infect a wide range of host plants, are seedborne, and are vectored by dorylaimid nematodes. The nematode *Xiphinema diversicaudatum* (Micoletzky) Thorne vectors ArMV, SLRV, and possibly CLRV. RRSV and TBRV strains are transmitted by different *Longidorus* species. These viruses are good immunogens and are easily identified in serological tests. However, for the most part, these viruses do not share common antigenic sites and hence are not closely related.

Disease Cycle and Epidemiology

Many tree species other than stone fruits, including shrubs, vegetables, and weeds, are natural hosts for nepoviruses. Weed hosts in particular could serve as reservoirs for virus acquisition by feeding nematode vectors. Seed transmissions have been demonstrated in a number of flowers, vegetables, and weeds. Natural *Prunus* hosts of nepoviruses include apricot, cherry, and peach for ArMV; sweet and sour cherries for CLRV; cherry and plum for RRSV; apricot, cherry, peach, and plum for SLRV; almond, cherry, and peach for TBRV; and Myrobalan plum rootstocks for MLRV. ArMV, CLRV, and SLRV have also been isolated from ornamental *Prunus*. However, only CLRV is known to spread from an infected stone fruit tree (*P. serotina*) into its seed and infect germinating seedlings.

Control

The most effective means of controlling nepoviruses is to prevent their introduction. Certification programs for domestic cultivars and quarantine testing of imported stone fruits are needed to ensure availability of virus-tested propagation material. In contaminated areas, site selection, soil fumigation, weed control, decontamination of farm equipment, and use of resistant or highly tolerant certified cultivars can minimize the impact of these pathogens.

Selected References

Harrison, B. D., and Murant, A. F. 1977. Nepovirus group. Descriptions of Plant Viruses, No. 185. Commonwealth Mycological Institute and Association of Applied Biologists, Kew, England. 4 pp.

Lamberti, F. 1981. Combating nematode vectors of plant viruses. Plant Dis. 65:113-117.

Lamberti, F., Taylor, C. E., and Seinhorst, J. W., eds. 1975. Nematode Vectors of Plant Viruses. Plenum Press, London. 460 pp.

Nemeth, M. 1986. Virus, Mycoplasma and Rickettsia Diseases of Fruit Trees. Martinus Nijhoff, Dordrecht, The Netherlands. 841 pp.

Taylor, C. E. 1980. Nematodes. Chapter 16 in: Vectors of Plant Pathogens. K. F. Harris and K. Maramorosch, eds. Academic Press, New York.

(Prepared by J. A. Foster)

Viruses and Pathogens Spread By Grafting (Natural Spread Apparent for Some Diseases)

Apple Chlorotic Leaf Spot Virus

Apple chlorotic leaf spot virus (ACLV) was described initially in apple and found later in all tree fruits that have been tested for the presence of this virus. It is one of the most common tree fruit viruses worldwide. Although many cultivars are symptomless hosts, ACLV can cause leaf symptoms in peach and fruit marking or graft-incompatibility diseases in the other stone fruits.

Symptoms

Symptom expression of ACLV is highly variable. It depends mostly on the strain of the virus and the specific reaction of the host cultivar. In peach, for example, leaf symptoms of dark green, sunken mottle (Plate 129) develop, but in most stone fruits, ACLV causes no foliage symptoms. However, fruit symptoms resembling apricot ring pox can be induced in apricot and plum. Also, fruit distortion can occur in cherries infected by certain strains of ACLV together with Prunus necrotic ringspot virus. Some virus strains cause a bark split in plum, whereas others induce symptoms of graft-incompatibility in apricot and peach.

Causal Agent

ACLV is a member of the closterovirus group and features flexuous rods approximately 720 nm long, containing a single-stranded RNA with a molecular weight of 2.5×10^6. The molecular weight of the coat protein subunit is 23,500. The virus can be purified by various means, but yields are moderate to poor, which probably reflects low concentration of the virus in the plant. The virus spreads primarily through infected plant material. All strains can be sap-transmitted to *Chenopodium quinoa* when nicotine-buffered extracts are prepared from recently expanded buds, petals, or young leaves. Occasional isolations from bark and fruit tissues have been reported.

Disease Cycle and Epidemiology

Virus transmission by vectors has not been reported, but natural spread by root graft has occurred in field-grown trees. Experimental transmission of ACLV from peach to peach by a contaminated budding knife has been reported in one case, but experimentation is lacking to determine whether such spread is common in the field. ACLV originally was described as a common latent virus of *Malus* spp. and was later found to occur in most stone fruits, including sweet cherry, sour cherry, peach, apricot, and plums. ACLV is found frequently in stone fruits that are routinely budded or grafted onto vegetatively propagated rootstocks. The virus has spread mostly through infected understocks or infected cultivars.

Control

ACLV can be controlled easily through use of virus-free understocks and budwood. Newly established plantings remain free of the virus wherever clean planting material has been used. Chemotherapy of ACLV-infected material with ribavirin is technically possible and relatively easy but is not economical under commercial conditions. Virus-free material can be obtained readily through heat treatment, selection of spontaneous virus-free shoots, or meristem culture followed by indexing.

Selected References

Desvignes, J. C., and Boye, R. 1988. Different diseases caused by the chlorotic leafspot virus on fruit trees. Acta Hortic. 235:31-38.

East Malling Research Station. 1967. Virus diseases of fruit trees. E. Malling Res. Stn. Annu. Rep. 1966, p. 35.

Guengerich, H. W., and Millikan, D. F. 1965. Root grafting, a potential source of error in apple indexing. Plant Dis. Rep. 49:39-41.

Hansen, A. J., and Gilmer, R. M. 1976. Apple chlorotic leafspot and other filamentous virus infections in *Prunus*. Pages 206-208 in: Virus Diseases and Noninfectious Disorders of Stone Fruits in North America. U.S. Dep. Agric. Agric. Handb. 437.

Hansen, A. J., and Lane, W. D. 1985. Elimination of apple chlorotic leafspot virus from apple shoot cultures by ribavirin. Plant Dis. 69:134-135.

Paul, F., and Dunez, J. 1988. Use of monoclonal antibodies for the detection of the apple chlorotic leafspot virus. Acta. Hortic. 235:49-56.

Smith, P. R. 1972. Transmission of apple chlorotic leafspot virus from peach to peach on the budding knife. Aust. Plant Pathol. Newsl. 1:12.

(Prepared by A. J. Hansen)

Apricot Ring Pox and Cherry Twisted Leaf

Apricot ring pox is known by various names, such as apricot ringspot and apricot pit pox, in apricot-growing areas west of the Rocky Mountains in North America. Based on results of reciprocal graft-transmissions, apricot ring pox is presumably caused by the same agent that causes cherry twisted leaf in sweet cherry, which has been reported from Oregon, Washington, and Idaho in the United States and British Columbia, Canada.

Symptoms

In apricot, the first leaves develop without symptoms. Vein-banding, chlorotic spots, streaks, and rings (the latter expand to form a flame pattern) develop on later-emerging leaves and increase in severity during the growing season. In midsummer, many of the spots and rings become necrotic and drop off, giving a shot-hole appearance (Plate 130). Another diagnostic symptom consists of dark purple discoloration of short sections (1–3 mm) of at least a few petioles in affected trees. Similar dark red spots can appear on current-season twigs. The most typical diagnostic feature of ring pox are the black pox or concentric rings (Plate 131), which can extend deep (i.e., forming a plug-like structure) into the flesh, reaching as far as the endocarp (seed) (Plate 132) but not affecting the integument (seed coat). The size and depth of these plugs and necrotic rings are extremely variable, ranging from a few millimeters to 2 cm. Fruit symptoms appear in a few fruits in one year and on all fruits in following years. Distribution of affected fruits in the tree can be very irregular. The apricot cultivars Tilton and Wenatchee Moorpark are most severely affected, whereas Royal, Reliable, and Sunglo are usually symptomless.

In sweet cherry, the most consistent symptom is a twist in the midrib or petiole of leaves (Plate 133) formed during the first third of the growing season. Parts of the midrib and the lateral veins may become necrotic. Leaf symptoms occur annually. All branches may be affected, or the disease may occur predominantly in one part of the tree or on one branch only. Affected shoots tend to be stunted, and the lower leaves may abscise after the first third of the growing season. Irregular fruit distortion often follows infection by certain strains of the causal agent. These irregular fruit symptoms are difficult to distinguish from those induced by tomato bushy stunt virus (described later in this section) or by green soldier (stink) bug (*Acrosternum hilaris* Say). Isolates that induce fruit symptoms also tend to induce gumboils and necrotic lesions on branches and on the trunk.

Causal Agent

The causal agent is unknown. Recently, a rod-shaped virus approximately 800 nm long has been transmitted to susceptible tobacco plants previously inoculated with extracts prepared from trees symptomatic for twisted leaf, ring pox, and pit pox diseases. Transmission studies are being conducted to determine relationships between these virus isolates and the stone fruit diseases. Also, in support of the suspected viral etiology, a high-molecular-weight (8.4×10^6) dsRNA was consistently recovered from sweet cherry trees with cherry twisted leaf disease.

Disease Cycle and Epidemiology

Both diseases spread locally in the field, but no vector has yet been identified. The only known natural hosts of ring pox and twisted leaf are chokecherry, hybrid plums (*Prunus salicina* × *P. simonii*), and sweet cherry. Chokecherry and plums develop very few symptoms after infection. The causal agent can be experimentally introduced into certain cultivars of almond, nectarine, Japanese plum, peach, sour cherry, desert peach (*P. andersonii* A. Gray), and Bessey cherry (*P. besseyi* L.H. Bailey). Abundance plum, Italian prune, American plum, and bitter cherry (*P. emarginata*) are immune to the agents of both diseases.

Control

With most outbreaks of ring pox and twisted leaf in Canada, infected chokecherry bushes have been found within 500 m of the diseased orchards. Removal of these source trees has prevented new infections. Once the disease is established in an orchard, it can spread readily from tree to tree. Immediate removal of infected orchard trees and use of virus-free nursery stock are recommended. Scion wood can be freed of the pathogen by heat treatment.

Selected References

Hansen, A. J. 1975. Differences between twisted leaf and tomato bushy stunt in sweet cherry. Acta Hortic. 44:55-57.

Hansen, A. J., and Cheney, P. W. 1976. Cherry twisted leaf. Pages 222-225 in: Virus Diseases and Noninfectious Disorders of Stone Fruits in North America. U.S. Dep. Agric. Agric. Handb. 437.

Hansen, A. J., Parish, C. L., and Pine, T. S. 1976. Apricot ring pox. Pages 45-49 in: Virus Diseases and Noninfectious Disorders of Stone Fruits in North America. U.S. Dep. Agric. Agric. Handb. 437.

James, D. 1992. Isolation of a high molecular weight dsRNA associated with twisted leaf disease in cherry. Can. J. Plant Pathol. 14:281-284.

Zhang, Y. P., Mink, G. I., Tiffany, M. G., and Howell, W. E. 1992. Isolation of viruses associated with cherry twisted leaf, apricot ring pox, and apricot diseases and their relationship to apple stem pitting virus. (Abstr.) Phytopathology 82:1149.

(Prepared by A. J. Hansen and G. I. Mink)

Cherry Black Canker

Cherry black canker disease of sweet cherry has been reported in the western United States, Canada, and Australia. Because of its limited occurrence, the disease is minor in importance.

The characteristic symptom is rough black cankers on the scaffold limbs and branches. The cankered areas begin as a swelling that later splits open to form thickened, ulcerlike wounds. These wounds enlarge to encircle the branch and kill the distal portion. Trees eventually decline and die. No fruit or leaf symptoms have been observed on otherwise symptomatic trees.

Cherry black canker is caused by a graft-transmissible agent. Graft-inoculated indicator trees require an incubation period of two to three years for symptoms to develop. Only cultivars of *Prunus avium* are known natural hosts of cherry black canker. Napoleon (Royal Ann) is a sensitive indicator of the disease.

Scion wood for propagation should be obtained from certified pathogen-free sources.

Selected Reference

Cheney, P. W., Blodgett, E. C., and Parish, C. L. 1976. Black canker. Pages 209-210 in: Virus Diseases and Noninfectious Disorders of Stone Fruits in North America. U.S. Dep. Agric. Agric. Handb. 437.

(Prepared by H. E. Waterworth)

Cherry Necrotic Rusty Mottle, Lambert Mottle, and Frogmore Virus Canker

The diseases necrotic rusty mottle (in the United States) and Lambert mottle (in Canada) have been reported from west of the Rocky Mountains and are considered to be caused by the same agent. The disease is common in parts of England, where it is known as Frogmore virus canker and where it was likely introduced by infected plant material from North America.

Symptoms

In affected cherry trees, brown, angular necrotic spots and slots appear on the leaves three to six weeks after full bloom (Plate 134). The necrotic areas either dry and dehisce or continue to expand until most of the leaf is affected. At this point, the leaves tend to turn yellow along the veins, and leaf abscission can occur. Symptoms vary from one cultivar to another; Lambert is one of the most severely affected, and Royal Ann is symptomless. Symptoms also vary with temperature; severe symptoms accompany cold springs. As the disease spreads in the tree, limbs and branches may be affected. Shallow necrotic areas and gum blisters develop on branches, especially on Bing, whereas general bark necrosis is more common on Lambert. Affected branches die back after a number of years, partly in response to a decrease in winterhardiness. In areas with occasional severe winters (–25°C or lower), infected trees can be winter-killed.

Causal Agent

The disease is caused by a graft-transmissible agent and can spread in a single year throughout an inoculated tree. The agent may be unrelated to that causing other diseases of the rusty mottle group. The proposed identity of necrotic rusty mottle, Lambert mottle, and Frogmore virus canker is based only on similarities in symptoms and host range.

Disease Cycle and Epidemiology

Field occurrence and patterns of spread indicate that the disease usually originates from symptomless infected trees and that a presumed vector spreads the agent within the orchard. The causal agent has never been found in native vegetation. In the field, the disease has been found on sweet cherry only. The cultivars Lambert and Sam are severely affected. Sour cherry, peach, apricot, Damson plum (*Prunus domestica* L. subsp. *insititia*), Nanking cherry, and mahaleb cherry can be infected experimentally.

Control

Use of virus-free budwood and understocks prevents the introduction of the disease into healthy orchards. Control by eradication of all symptomless and visibly affected trees has been successful wherever it has been tried. Care should be taken to test all trees in an orchard, because some major cultivars, including Bing, can be symptomless carriers. The disease is readily eliminated from affected budwood by thermotherapy.

Selected References

Moore, D. L., and Cameron, H. R. 1983. A nucleic acid associated with necrotic rusty mottle diseased cherries. Acta Hortic. 130:80.

Wadley, B. N., and Nyland, G. 1976. Rusty mottle group. Pages 242-249 in: Virus Diseases and Noninfectious Disorders of Stone Fruits in North America. U.S. Dep. Agric. Agric. Handb. 437.

(Prepared by A. J. Hansen)

Cherry Rusty Mottle

Rusty mottle disease was reported first in Washington State in 1940. Since then it has been found in Oregon, Idaho, Montana, and British Columbia but has not been reported outside North America. The disease occurs in several orchards in the Pacific Northwest, where it apparently causes little economic damage.

Symptoms

Leaf symptoms first appear as chlorotic spots or rings on basal leaves four to five weeks after full bloom (Plate 135). Subsequently, chlorotic rings develop on expanded leaves of rapidly growing shoots located within the canopy of the tree. Just before harvest, many older leaves on main branches become bright yellow and then develop a mottle of brown, orange, green, and yellow shades. Many of the mottled leaves abscise. Considerable defoliation can occur from midsummer through early fall. Chronically affected trees exhibit weak growth and decline progressively over time. Fruit on newly affected trees ripen normally, whereas those on chronically affected trees tend to be small and have less color than fruit on healthy trees.

Causal Agent

The causal agent has not been identified, but it is graft-transmissible. All sweet and sour cherry cultivars that were graft-inoculated developed disease symptoms. Other susceptible hosts were apricot, peach, chokecherry, Damson plum, Nanking cherry, and mahaleb cherry.

Disease Cycle and Epidemiology

The disease spreads from infected trees to neighboring healthy ones in sweet cherry orchards. However, the high incidence of disease in commercial orchards results from propagations using infected plant material.

Control

Tree removal is the only eradicative control. However, growers seldom remove infected trees, because the disease causes little economic damage.

Selected Reference

Wadley, B. M., and Nyland, G. 1976. Rusty mottle group. Pages 242-249 in: Virus Diseases and Noninfectious Disorders of Stone Fruits in North America. U.S. Dep. Agric. Agric. Handb. 437.

(Prepared by G. I. Mink)

Cherry Short Stem

Short stem of sweet cherry was found near Flathead Lake, Montana, in 1958 and near Hood River, Oregon, in 1962. A year after inoculation, affected leaves develop constriction and necrosis of the midribs, resulting in twisting of some spur leaves. Initial fruit symptoms include dappling, rough surface, and irregular shape. The most prominent feature of this disease is the normally developed fruit with shortened stems (Plate 136), which appear in the second year of infection and thereafter.

The agent is transmitted by budding and grafting. All sweet cherry cultivars are susceptible to the pathogen. In addition, graft-inoculated trees of Abundance plum (*Prunus salicina*) and seedlings of Amur chokecherry and Chinese wild peach develop twisted leaf symptoms. Also, Early Red Haven peach produced pink-yellow fruit with an uneven surface and firm-textured flesh even under postharvest storage. Symptomless hosts include Drake almond, Riland and Tilton apricot, Elberta, Fairhaven, and J. H. Hale peach, and seedlings of western sand cherry, pin cherry, and wild black cherry. Nonhosts were chokecherry, Nanking cherry, desert cherry, American plum, Italian prune, and Shiro plum (*Prunus* hybrid).

All diseased trees should be removed, and only certified trees should be replanted.

Selected References

Afanasiev, M. M., and Hamilton, R. I. 1965. Short stem virosis of sweet cherries. Plant Dis. Rep. 41:253-255.

Cheney, P. W., Parish, C. L., and Coyier, D. L. 1970. Host range studies of the cherry short stem virus from Oregon. Plant Dis. Rep. 54:651-655.

Li, T. S. C. 1994. Early Red Haven peach, a host of cherry spur cherry virus. Fruit Var. J. 48(3):184-185.

Parish, C. L., and Cheney, P. W. 1976. Short stem. Pages 250-251 in: Virus Diseases and Noninfectious Disorders of Stone Fruits in North America. U.S. Dep. Agric. Agric. Handb. 437.

(Prepared by C. L. Parish)

Cherry Stem Pitting

Originally described in 1977, cherry stem pitting is characterized by stem pitting and decline of sweet cherry trees in California. A study completed in 1983 failed to show the presence of tomato ringspot virus (Prunus stem pitting) or cherry leafroll virus in diseased trees.

Symptoms

The initial spring symptom is delayed budbreak. Leaves are small and more erect than those on healthy limbs or trees. The lower portions of affected trees are often bare. Occasionally, only a single scaffold branch may be involved. Diseased fruit are unmarketable because they lack adequate size and color and are borne on short pedicels. Bark tissues are thickened, and the wood beneath the bark of some scion cultivars exhibits pits and grooves (Plate 137). Stem pitting symptoms occur in greater intensity at or near the soil line and become less severe in the upper portion of the trunk and scaffold branches. Severe pitting occurs on mazzard rootstocks but not on mahaleb rootstocks. Affected trees decline rapidly and die.

Causal Agent

Cherry stem pitting is caused by a graft-transmissible agent of unknown etiology. Bing trees are often severely affected, whereas Napoleon is a symptomless host. In affected Bing cherry orchards, Bing scions on either mahaleb or mazzard rootstocks develop symptoms. The disease agent was transmitted experimentally to Stockton Morello, Nanking cherry, and Lovell peach.

Disease Cycle and Epidemiology

Rapid disease spread (incidence up to 44%) has occurred in several mature cherry orchards in California. Replants of young cherry trees in sites where diseased trees were present previously often become infected after a year or more. The observed spatial patterns of diseased trees and development of stem pitting symptoms suggest that the disease agent is soilborne.

Control

In cherry replant trials that are still in progress, several Bing scions grown on mahaleb rootstock were diseased, whereas all Bing scions grown on Colt rootstock (some exposed for four seasons) were healthy (J. K. Uyemoto, *unpublished*). Colt may prove to be resistant to the disease agent and perhaps provide effective control in previously diseased orchards.

Selected References

Hoy, J. W. 1983. Studies on the nature of the prune brownline and sweet cherry stem pitting diseases and epidemiology and control of prune brownline. Ph.D. thesis. University of California, Davis. 69 pp.

Mircetich, S. M., Moller, W. J., and Nyland, G. 1977. Prunus stem pitting in sweet cherry and other stone fruit trees in California. Plant Dis. Rep. 61:931-935.

Uyemoto, J. K., Luhn, C. F., and Griesbach, J. A. 1987. A severe outbreak of stem pitting in young trees of sweet cherry. (Abstr.) Phytopathology 77:1242.

(Prepared by J. K. Uyemoto)

Peach Mottle

Peach mottle was observed in a few Elberta peach trees in the state of Idaho (around 1940) and in Italy. It is limited in occurrence and is not important economically.

Affected leaves are slightly smaller than healthy ones and exhibit a yellow mottle; leaf margins are turned upward and puckered (Plate 138). The tree canopy is lighter in color and ragged in appearance, but these symptoms fade by the middle of the growing season. Infection of young trees results in death.

Peach mottle is caused by an unidentified graft-transmissible agent. Indicator trees inoculated in the fall often produce symptoms the following spring. The peach cultivar Elberta is the only known natural host of the disease agent. Other susceptible hosts include J. H. Hale peach, sweet cherry cultivars Black Republican, Bing, Lambert, and Napoleon, and Montmorency sour cherry. Graft-inoculations to cherries produced symptoms of leaf mottle, leaf and shoot necrosis, and tree stunt (in sweet cherry) or mottled leaf, twig dieback, and gum pockets (in sour cherry).

Removal of infected trees and propagation of only pathogen-free material are the practices used to control peach mottle.

Selected Reference

Helton, A. W. 1976. Peach mottle. Pages 71-72 in: Virus Diseases and Noninfectious Disorders of Stone Fruits in North America. U.S. Dep. Agric. Agric. Handb. 437.

(Prepared by H. E. Waterworth)

Peach Wart

Peach wart was recognized first in Idaho (around 1938) and later in Oregon, Washington, California, and New Mexico in the United States; it also was found in Italy, Turkey, and possibly France. The disease is considered rare.

On sensitive peach cultivars, wartlike outgrowths develop on the fruit surfaces or in a restricted region around the stylar end or along the suture ("crease wart") (Plate 139) of the young fruit. Affected fruit may be small and misshapen. The tissues beneath the affected area are coarse-textured and contain gum pockets. With these exceptions, the disease appears to have no effect on the vegetative growth and development of infected peach trees.

Peach wart is caused by an unidentified graft-transmissible agent. Peach is a natural host of the pathogen. With a bud-chip, a contact period as short as four to five days was sufficient to transmit the agent to a healthy tree. The agent is not seedborne but was experimentally transmitted to nectarines, sweet and sour cherries, flowering cherries, apricots, Nanking cherry, and Shiro plum. Although only fruit symptoms were induced on several peach and nectarine cultivars, necrosis of shoot tips was induced on Shiro plum following three to four weeks of incubation.

Diseased trees should be destroyed promptly. Pathogen-free propagation material helps to prevent introduction of the pathogen into orchards.

Selected References

Blodgett, E. C. 1976. Peach wart. Pages 88-90 in: Virus Diseases and Noninfectious Disorders of Stone Fruits in North America. U.S. Dep. Agric. Agric. Handb. 437.

Fridlund, P. R. 1967. Symptoms of the peach wart virus in some *Prunus* species and peach and nectarine varieties. Plant Dis. Rep. 51:42-43.

Lazar, A. C., and Fridlund, P. R. 1967. Shiro plum: A rapid greenhouse indicator for the peach wart virus. Plant Dis. Rep. 51:831-833.

(Prepared by J. K. Uyemoto)

Plum and Peach Dapple Fruit Viroid

Dapple fruit diseases of plum and peach recently were described in Japan. Viroids related to hop stunt viroid have been associated with both diseases. The diseases are not known to occur outside Japan. Because the viroid can remain symptomless in some trees, the disease could spread from Japan through overseas shipments of uncertified budwood.

Symptoms have been observed on several cultivars of plums and peaches. Clear reddish and chlorotic blotches appear on diseased fruit as it begins to color (Plate 140). The blotches become prominent on mature fruit of plums such as Santa Rosa and several peach cultivars. Fruit of Soldam plum turn yellowish red, and the fruit surface appears waxy. No symptoms occur in the flesh of fruit or on leaves.

The viroids isolated from dappled plum and peach trees are similar. Both are considered to be strains of hop stunt viroid, which is found in both hop and grapevine. No information is available on natural spread of these agents. The viroids have been transmitted mechanically to cucumber.

Trees should be propagated from viroid-free budwood.

(Prepared by G. I. Mink)

Sour Cherry Green Ring Mottle Virus

Green ring mottle was reported first on Montmorency sour cherry (*Prunus cerasus* L.) in Michigan in 1937. It has since been recognized as an important virus disease worldwide.

Symptoms

Symptoms are best expressed on Montmorency sour cherry and Kwanzan and Shirofugen flowering cherry. Leaves of infected sour cherry trees show yellow mottling with green islands or ringlike bands and constricting chlorosis along the veins and midrib (Plate 141). Affected fruit are misshapen and bitter and may exhibit necrotic pits or dimples on the surface. Kwanzan and Shirofugen flowering cherry trees develop necrosis of the petiole, midrib, and lateral veins that results in twisting and curling of the leaves (Plate 142). Longitudinal cracks and fissures may occur in the bark (rough bark symptoms). Aphid feeding can mimic the leaf epinasty caused by green ring mottle virus. On flowering cherry cultivars, visible symptoms develop six weeks after inoculation if the ambient temperature is between 18 and 20°C; otherwise, symptoms may be masked. In Duke cherry (*P. cerasus* × *P. avium* 'Late Duke', 'May Duke', and 'Reine Hortense'), the virus incites a green and yellow leaf mottle.

Causal Agent

Green ring mottle virus is a member of the closterovirus group. Virus particles purified directly from infected cherry and peach tissues measured 1,000–2,000 nm in length and 5–6 nm in width. Fibrillar inclusions were present in infected cells. The virions were composed of a single-stranded RNA and a protein subunit with molecular weights of 2.5×10^6 and 25,000, respectively. Natural hosts of the virus include sour cherries and Japanese flowering cherries. Duke cherries are symptomatic hosts; apricot, peach, sweet cherry, and mahaleb cherry are symptomless hosts. The virus is sensitive to thermotherapy at 38°C for six weeks.

Disease Cycle and Epidemiology

The disease has been observed to spread slowly under field conditions; newly infected trees usually occur near known diseased trees, suggesting that root grafts may be involved. No vector has been identified. In new plantings, disease is established through use of infected plant material.

Control

Diseased trees should be removed and replaced with virus-tested stock.

Selected References

Bar-Joseph, M., and Murant, A. F. 1982. Closterovirus group. Descriptions of Plant Viruses, No. 260. Commonwealth Mycological Institute and Association of Applied Biologists, Kew, England. 3 pp.

Haufler, K. Z., Aref, N. M., and Ramsdell, D. C. 1989. Purification, serology, and some properties of a mechanically transmissible virus associated with green ring mottle disease in peach and cherry. Phytopathology 79:451-456.

Parker, K. G., Fridlund, P. R., and Gilmer, R. M. 1976. Green ring mottle. Pages 193-199 in: Virus Diseases and Noninfectious Disorders of Stone Fruits in North America. U.S. Dep. Agric. Agric. Handb. 437.

(Prepared by D. C. Ramsdell)

Sour Cherry Pink Fruit

Cherry pink fruit was first found in Washington State in 1934 on Montmorency sour cherry (*Prunus cerasus* L.). In the 1940s and 1950s, this disease was considered second in importance only to brown rot (caused by *Monilinia laxa*) in causing yield losses in Washington. The disease is now rare, due to a paucity of Montmorency orchards resulting from economic considerations in Washington State.

Symptoms

The first visible symptom is delayed bloom. Affected leaves roll slightly toward the midrib and thereby mimic symptoms associated with drought stress. Tree growth is suppressed. Fruit symptoms begin to appear about two weeks before harvest. Fruits are slightly smaller than healthy fruit and are pinkish to pinkish brown (Plate 143), in contrast to the bright red color of healthy fruit. Necrotic areas develop in the fruit flesh. Normal and diseased fruit may form side by side on the same spur.

Disease severity depends on the scion-rootstock combination. Montmorency scions on mahaleb rootstock express slight symptoms during the second year after inoculation and are in full remission thereafter. In contrast, Montmorency scions grafted on mazzard rootstock or self-rooted Montmorency trees develop intense disease symptoms affecting nearly all of the fruit.

Causal Agent

Sour cherry pink fruit is caused by a graft-transmissible agent. Montmorency cherry trees exhibit symptoms 15–27 months after being grafted with diseased inoculum. Sweet cherry (i.e., Bing and Lambert) is a symptomless host. Immune hosts include Italian prune, bitter cherry, mahaleb cherry, and peach. Diseased fruit of Montmorency are similar in appearance to those described for X-disease of sweet cherry, but repeated cross-inoculations onto sweet and sour cherry indicator trees failed to demonstrate a relationship between the two disease agents. The diseases are considered biologically distinct.

Control

Only disease-free propagation material should be used when new orchards are established.

Selected Reference

Cheney, P. W., Parish, C. L., and Johnson, F. 1976. Pink fruit. Pages 200-201 in: Virus Diseases and Noninfectious Disorders of Stone Fruits in North America. U.S. Dep. Agric. Agric. Handb. 437.

(Prepared by C. L. Parish)

Spur Cherry

Spur cherry was discovered first in a Yakima, Washington, orchard in an eight-year-old Bing cherry tree bearing normal fruit that was initially considered a dwarf mutant of Bing. Commercialization of this spur-type Bing was vigorously pursued. Later, however, graft-transmissibility of the condition was demonstrated. The disease may be widely distributed, because nursery stock was sold as Dwarf Bing cherry, and because diseased budwood was freely sent out to various states and other areas of the world for experimental purposes.

Symptoms

Certain sweet cherry cultivars exhibit severe rosetting, epinasty, and bark necrosis in new growth. Among other cultivars that were experimentally infected, some grew in a spiral fashion, some were dwarfed, and some had no symptoms. No fruit symptom has been observed. Spur cherry causes moderate to severe stunting of shoot growth in Bing, Sam, Van, and other sweet cherry cultivars. In contrast, susceptible cultivars such as Black Tartarian, Chinook, Lambert, Long Stem, and Bing and sour cherry cultivars do not show symptoms when infected. Graft-inoculated Tilton and Riland apricots, however, displayed symptoms similar to those of apricot ring pox disease. Symptomless hosts include J. H. Hale peach, Kwanzan flowering cherry, and seedlings of Nanking cherry, wild black cherry, and Manchurian cherry (*Prunus maackii*).

Causal Agent

Spur cherry is caused by a graft-transmissible agent that has not been identified.

Control

Propagation of trees with budwood from virus-indexed mother stock controls this disease.

Selected References

Blodgett, E. C., and Aichele, M. D. 1976. Spur cherry. Pages 252-253 in: Virus Diseases and Noninfectious Disorders of Stone Fruits in North America. U.S. Dep. Agric. Agric. Handb. 437.

Blodgett, E. C., Jenkins, P., and Aichele, M. D. 1965. A spur type growth of Bing cherry caused by virus. Plant Dis. Rep. 49:910.

Cheney, P. W., Aichele, M. D., Parish, C. L., and Blodgett, E. C. 1969. Host range of the spur cherry virus. Plant Dis. Rep. 53:984-988.

(Prepared by M. D. Aichele)

Tomato Bushy Stunt Virus

Tomato bushy stunt virus (TBSV) has been reported from many crops in the Americas, Europe, and North Africa. Natural infections on sweet cherry have been found in Germany and Switzerland; Ontario and British Columbia, Canada; and Idaho, United States. Infected plums (*Prunus domestica* and *P. salicina*) and sour cherry were found near Prague, Czech Republic. Apricots were found also to be naturally infected in British Columbia. The disease has been described under the synonyms gumming, distortion, and pitting; detrimental canker (sensu Blattny and Schmid); xylem aberration; and target spot. Because of its destructive effects on infected trees, TBSV is an economically important pathogen.

Symptoms

The most obvious symptom of TBSV infection in cherry and plum is severely and irregularly distorted fruits. On Lambert sweet cherry, a necrotic central spot is surrounded by one or two necrotic circles, giving the appearance of a target spot. Necrotic fruit symptoms (Plate 144) are accompanied by sporadic necrotic lesions or spots on or near the midrib of leaves. Affected leaves and fruit pedicels often display a sharp bend at the point of necrosis (Plate 144), while petals become necrotic and misshapen. Necrosis of one-year-old twigs leads to twig-bending at the affected point. Some isolates of TBSV regularly induce necrosis and large gum-filled blisters on major limbs as well as pitting symptoms on the wood of the trunk of cherry trees. Only parts of affected trees may show symptoms, and the virus is recovered only from symptomatic tissues, especially at the advancing edge of necrotic lesions. Trees infested by the green soldier (stink) bug (*Acrosternum hilaris* Say) or infected with some strains of apple chlorotic leaf spot virus can display similarly distorted fruits, but they lack the leaf and twig symptoms characteristic of TBSV infections.

Causal Agent

TBSV belongs to the tombusvirus group, with isometric particles 30 nm in diameter. The genome is a single-stranded, positive sense RNA with a molecular weight of 1.5×10^6 and a protein shell assembled with 180 subunits of a single protein species with a molecular weight of 40,000. The virus reaches high concentrations in plant cells during active replication, reaching dilution endpoints of 10^{-6} in *Chenopodium quinoa.* The thermal inactivation point is between 80 and 90°C. Indexing for TBSV is done by sap transmission from symptomatic flowers, fruits, young leaves, or twigs to *C. quinoa* or other herbaceous hosts. Serological assays with homologous strains have been developed. Indexing on woody hosts is not recommended, since the virus is distributed very irregularly in infected trees. The virus usually induces severe necrotic lesions on the inoculated leaves of herbaceous hosts. Symptoms can appear within 48 hr. TBSV can be distinguished from tobacco necrosis virus by the systemic symptoms that TBSV induces in *Nicotiana clevelandii* and by serology.

Disease Cycle and Epidemiology

TBSV is soilborne, has no known vector, and is probably primarily a root-infecting virus that rarely becomes systemic in aboveground parts of fruit trees. Natural hosts of TBSV include sweet cherry, apple, privet (*Ligustrum vulgare* L.), grapevine, and many cultivated and native herbaceous plants.

Control

TBSV-free budwood can easily be grown from virus-free buds of partially infected trees. No specific control measures have been tested.

Selected References

Allen, W. R., and Davidson, T. R. 1967. Tomato bushy stunt virus from *Prunus avium* L. 1. Field studies and virus characterization. Can. J. Bot. 45:2375-2382.

Hansen, A. J. 1975. Differences between twisted leaf and tomato bushy stunt in sweet cherry. Acta Hortic. 44:55-57.

Koenig, R., and Kunze, L. 1982. Identification of tombus virus isolates from cherry in southern Germany as petunia asteroid mosaic virus. Phytopathol. Z. 103:361-368.

Lott, T. B., Keane, F. W. L., and May, J. 1962. Gumming, distortion, and pitting in cherry and apricot. Can. Plant Dis. Surv. 42:229-232.

Novak, J. B., and Lanzová, J. 1977. Identification of tomato bushy stunt virus in cherry and plum trees showing fruit pitting symptoms. Biol. Plant. (Prague) 19:234-237.

Novak, J. B., and Lanzová, J. 1980. Some diseases of fruit trees in which the tomato bushy stunt virus occurs and new natural hosts of this virus. Acta Phytopathol. 15:323-327.

Schmid, G. 1968. Investigation on detrimental canker of sweet cherries. Eur. Symp. Tree Fruit Virus Dis. 7th. Tagungsber. Akad. Landwirtschattswiss. DDR 37:155-163.

(Prepared by A. J. Hansen)

Other Graft-Transmitted Pathogens

Other virus and viruslike diseases of stone fruits have been reported (Table 4). Even though graft transmission has demonstrated the presence of at least one infectious agent for each of these diseases, the pathogen or pathogens for many are unknown. Currently, the only reliable method for detecting these infectious agents is by grafting candidate shoot buds or bark patches onto sensitive indicator hosts. Symptom development, and therefore diagnosis, requires an incubation period ranging from a month to two or more years.

The characterization and distribution of many of the diseases and pathogens listed in Table 4 are incomplete. Comparing disease agents in distant countries with those in the United States is difficult, because variations among pathogen strains, use of different indicator hosts, and climatic influences affect disease expression. Causal agents or combinations thereof may produce similar symptoms, but strains of the same pathogen may cause very different symptoms under similar environmental conditions. As research elucidates properties for what now appears to be a diverse array of pathogens, current perceptions of these diseases will change.

The limited information available indicates that most diseases listed in Table 4 may be of minor importance in the places where they are found. Only a few are suspected to spread naturally. Even so, the potential economic impact of many of these diseases may change if an efficient vector is introduced, if the pathogen spreads into another area inhabited by an efficient vector, if a more virulent pathogen or efficient vector is selected, or if transmission into a more sensitive cultivar or a different crop plant is successful.

The most effective control for pathogens that do not have natural vectors is to use pathogen-tested propagation material. Topworking existing trees with certified scions or grafting certified trees with uncertified scions is discouraged. International regulation of the distribution of propagation material from pathogen-tested sources by governmental agencies plays a vital role in limiting further spread of these potentially destructive pathogens.

Selected References

Gilmer, R. M., Moore, J. D., Nyland, G., Welsh, M. F., and Pine, T. S., eds. 1976. Virus Diseases and Noninfectious Disorders of Stone Fruits in North America. U.S. Dep. Agric. Agric. Handb. 437. 433 pp.

Nemeth, M. 1986. Virus, Mycoplasma and Rickettsia Diseases of Fruit Trees. Martinus Nijhoff, Dordrecht, The Netherlands. 841 pp.

Wood, G. A. 1979. Virus and viruslike diseases of pome fruits and stone fruits in New Zealand. N.Z. Dep. Sci. Ind. Res. Bull. 226. 87 pp.

(Prepared by J. A. Foster)

TABLE 4. Other Viruslike Pathogens Reported in Stone Fruits[a]

Pathogen	Hosts	Natural Symptoms	Transmission[b]	Distribution
Almond (infectious) bud failure agent; ?PNRSV	Almond, peach	Erect leaves, bud failure, bare branches, shoot dieback, deformed fruit, lower yield	B, G	USA (California)
Apricot chlorotic leaf mottle agent	Apricot	Chlorotic leaf spots & blotches, stunted growth	B, G	?Greece, ?India, New Zealand
Apricot deformation mosaic virus	Apricot	Deformed leaves, stunted growth, small & less fruit	B, G, Sa	Former Soviet Union
Apricot Moorpark mottle agent	Apricot	Light green leaf areas, small & distorted fruit	B, G	New Zealand (infected trees destroyed)
Apricot pucker leaf agent	Apricot	Chlorotic leaf flecks, puckered & tattered leaves, fruit spots	B, G	USA (Utah)
Apricot stone pitting agent	Apricot	Ring spots & veinbanding in leaves, distorted fruit	B, G	New Zealand (infected trees destroyed), ?Italy
Amasya cherry disease agent	Sweet cherry	Leaf spots, defoliation, small & tasteless fruit, shoot dieback, tree death	B, G	Turkey
Cherry freckle fruit agent	Sweet cherry	Small, deformed, flecked fruit	B, G	USA (Oregon)
Cherry Hungarian rasp leaf virus	Sweet cherry	Deformed, discolored leaves, leaf enations, small & discolored fruit	B, G, Sa	Bulgaria, Germany, Hungary, ?former Soviet Union
Cherry line pattern & leaf curl virus	Sweet cherry	Chlorotic lines in leaves, deformed leaves in cherry, peach fruit spots	B, G, Sa	Former Czechoslovakia, Hungary
Cherry rough bark agent	Sweet cherry	Bark blisters, gumming	B, G	USA (California)
Cherry rough fruit agent	Sweet cherry	Fruit spots, crinkled & deformed fruit	B, G	?Bulgaria, Iran, USA (Utah, infected trees destroyed)
Cherry rusty spot agent	Sweet cherry	Leaf spots, shot holes; curled, discolored leaves	B, G	New Zealand
Cherry xylem aberration agent	Apricot, cherry	Distorted leaves, fruit, & shoots; cankers; gumming; wood pitting	B, G	Canada (British Columbia)
European cherry rusty mottle agent	Sweet cherry	Leaf mottling, lower yield	B, G	Europe, ?South Africa
Sour cherry rusty splitting agent	Cherry	Bark splitting, gumming	B, G	USA (Oregon)
Sour cherry gummosis agent	Cherry	Tip dieback, cankers & gumming in sour cherry	B, G	USA (Oregon, Washington)
Sour cherry line pattern agent	Sour cherry	Spots & veinbanding in leaves	B, G	Former Czechoslovakia
Peach asteroid spot agent	Many stone fruits	Leaf spots, discoloration & defoliation, deformed fruit & twig dieback in cherry	B, G	?Europe, ?Israel, Mexico, ?Morocco, western USA
Peach bark & wood grooving agent	Peach	Branch & trunk grooves, weakened tree	B, G	?Greece, USA (Michigan)
Peach blotch agent	Peach	Chlorotic leaf areas	B, G	Canada, Italy, USA (California), ?former Yugoslavia
Peach calico agent	Peach, nectarine	Chlorotic to white spots in leaves, shoots, and fruit	B, G	Australia, ?Canada, Europe, New Zealand, USA
Peach chlorosis agent	Peach	Curling & chlorosis in leaves, shoot proliferation, stunted growth	B, G	France
Peach chlorotic spot agent	Nectarine	Leaf spots & distorted leaves in peach	B, G	New Zealand
Peach enation virus	Peach	Leaf enations & distortion	B, G, Sa	?Italy, Japan
Peach latent mosaic agent (viroid)	Peach	Deformed leaves; spotted, small, deformed fruit; bud and shoot necrosis	B, G, A	France, ?USA
Peach leaf necrosis agent	Plum	Leaf spots, stunted shoots & dieback in peach	B, G	Germany
Peach oil blotch agent	Peach, plum	Leaf blotches in peach	B, G	Japan

(continued on next page)

TABLE 4 *(continued)*

Pathogen	Hosts	Natural Symptoms	Transmission[b]	Distribution
Peach purple mosaic agent	Peach	Leaf flecks, tree decline, lower yield	B, G	Bulgaria, ?former Czechoslovakia, Romania, ?former Soviet Union
Peach pseudo stunt agent	Plum	Stunted growth in peach	B, G	Germany
Peach seedling chlorosis agent	Sweet cherry, plum	Chlorotic leaf areas in peach	B, G	New Zealand
Peach star mosaic agent	Peach	Chlorotic leaf spots, leaf distortion	B, G	Japan
Peach stubby twig agent	Peach, nectarine	Chlorotic & deformed leaves, thick twigs, bud failure, small fruit	B, G	?Italy, ?Turkey, USA (California, Missouri, Oregon)
Peach yellow mosaic agent	Peach	Leaf mosaic, distorted leaves	B, G	?Italy, Japan
Peach yellow mottle agent	Apricot	Mottle, ring spots, & line patterns in peach leaves	B, G	New Zealand
Krikon stem necrosis agent	Plum	Leaf spots, mottle, reduced growth, shoot necrosis, dieback	B, G, Se	France
Plum fruit crinkle agent	Plum	Spotted, wrinkled, distorted fruit; leaf line patterns; stunted growth	B, G	New Zealand (infected trees destroyed)
Plum mottle leaf agent	Plum	Leaf mottle, distorted leaves, stunted growth, lower yield	B, G	New Zealand
Plum ochre mosaic agent	Plum	Leaf spots, stunted growth	B, G, A	Former Czechoslovakia
Prune diamond canker agent	Prune	Trunk & branch cankers, rough bark, tree decline	B, G	?Chile, USA (California)

[a] A question mark preceding an entry indicates an unproven association only.
[b] B = budding, G = grafting, Sa = sap, Se = seed, A = aphid(s).

Part II. Noninfectious Disorders

Genetic and Physiological Disorders

Apricot Gumboil

Gumboil, reported in 1959, was found in scattered trees of Royal and Tilton apricots in several counties in California. It is of little importance economically. Only the apricot cultivars Royal and Tilton are known to develop symptoms.

Severely symptomatic trees were more than 10 years old. Affected trees were smaller and exhibited shoot dieback and less overall vegetative growth than healthy trees. The trunk and scaffold branches appeared rough with bark cracks. Pockets of necrotic tissues and accumulations of dark-colored gum (named gumboil) are present in inner bark tissues. Gum pockets extend into the xylem tissues of branches more than three years old.

Apricot gumboil is spread by vegetative propagation but is not systemically transmitted to healthy trees by budding or grafting. The rootstock appears to influence onset of symptoms; gumboil develops in trees grown on Myrobalan 29C and on seedlings of Myrobalan and apricot, but not on peach seedlings.

Use of healthy propagation material is the most appropriate practice to control apricot gumboil.

Selected Reference

Nyland, G. 1976. Apricot gumboil. Pages 261-264 in: Virus Diseases and Noninfectious Disorders of Stone Fruits in North America. U.S. Dep. Agric. Agric. Handb. 437.

(Prepared by H. E. Waterworth)

Plum Chlorotic Fleck and Peach Yellow Spot

Chlorotic fleck was first observed in New York (around 1941), where an incidence of 90% occurred in some Myrobalan nursery stock. Later, in 1952, yellow spot of peach was described in California. A similar leaf flecking of plum was observed in Germany in 1957.

Symptoms

In New York, symptoms of chlorotic fleck are noticeable in late summer (August–September) on one- and two-year-old Myrobalan seedlings. Circular, chlorotic spots and flecks are formed along the midrib and lateral veins. If flecks are abundant, the leaves become distorted and twisted. Severely affected seedlings are stunted. Oval, reddish flecks in the young bark are associated with this disorder.

In California, Lovell peach seedlings exhibit chlorosis along the primary and tertiary leaf veins. In severely affected leaves, the chlorotic areas coalesce to form a mosaic pattern. Myrobalan and Brompton plums and Lovell, Salberta, and J. H. Hale peaches are sensitive cultivars.

Cause

Both disorders are caused by a salivary toxin produced by the eriophyid mite *Vasates fockeui* (Nalepa & Trouessart). A mite infestation of very young seedlings of plum or peach will produce symptoms on leaves and shoots in 10–14 days. A feeding exposure of 24 hours is sufficient to induce yellow spot symptoms, which develop after two weeks. Mature leaves infested with mites remain free of chlorotic fleck or yellow spot symptoms.

The mites overwinter as diapausing females between bud scales of host plants and move immediately onto young leaves as they emerge in the spring. The reproductive cycle may be as short as one week during summer months, when mite populations may exceed 1,000 per leaf.

Control

The mite is easily controlled by timely use of various acaricides registered for use on stone fruits.

Selected References

Gilmer, R. M., and McEwen, F. L. 1958. Chlorotic fleck, an eriophyid mite injury of Myrobalan plum. J. Econ. Entomol. 51:335-337.

Oldfield, G. N. 1984. Evidence for conspecificity of *Aculus cornutus*, and *A. fockeui* (Acari: Eriophyidae), rust mites of *Prunus* fruit trees. Ann. Entomol. Soc. Am. 77:564-567.

Wilson, N. S., and Cochran, L. C. 1952. Yellow spot, an eriophyid mite injury on peach. Phytopathology 42:443-447.

(Prepared by G. N. Oldfield)

Plum Rusty Blotch

Plum rusty blotch was reported initially in California (1960), was later found in all plant-growing areas in the state, and is economically important. The disorder has also been reported in Argentina and Israel.

Chlorosis appears along the margins of leaves on affected trees (Plate 145). Within a few weeks, the chlorotic areas may turn reddish brown, and many small spots appear in the discolored areas. These spots may grow larger, but usually they become necrotic and drop out, giving the leaf a shot-hole appearance. Affected leaves are much smaller than normal. On symptomatic branches, only a few blossoms are produced, and these do not set fruit. Many cultivars of Japanese plums and hybrids of *Prunus salicina* Lindl. display leaf symptoms.

The disorder may be genetic in nature, because it was not graft-transmitted to healthy indicator trees.

Scion buds should be collected only from source trees that are known to produce quality trees of high fruit-yielding capacity.

Selected References

Pine, T. S. 1965. Plum rusty blotch in California. Plant Dis. Rep. 49:109-110.

Pine, T. S. 1976. Plum rusty blotch. Pages 294-296 in: Virus Diseases and Noninfectious Disorders of Stone Fruits in North America. U.S. Dep. Agric. Agric. Handb. 437.

(Prepared by H. E. Waterworth)

Prune Crinkle Leaf

Prune crinkle leaf was widespread in prune and plum nursery stock grown in Washington State during the 1960s. Now, increased grower awareness and a certification program have reduced the incidence of prune crinkle leaf considerably.

Affected trees produce leaves that are deformed and have uneven margins. Symptomatic leaves are small and light green with scattered chlorotic areas; they appear marbled in texture. Crinkle leaf trees are dwarfed and fail to produce mature fruit. The disorder was observed in Italian prune and in other prune cultivars such as Emily, Merton, and Hungarian.

Symptoms develop naturally in prune seedlings. Repeated attempts to graft-transmit crinkle leaf to healthy trees or top-work symptomatic trees with healthy scions have always failed to demonstrate transmission to the healthy indicator shoots. Hence, crinkle leaf is considered a genetic disorder.

Since this disorder is bud-perpetuated, a rigorous selection and propagation program for healthy budwood source trees should be implemented.

Selected Reference

Blodgett, E. C. 1976. Prune crinkle leaf. Pages 297-298 in: Virus Diseases and Noninfectious Disorders of Stone Fruits in North America. U.S. Dep. Agric. Agric. Handb. 437.

(Prepared by M. D. Aichele)

Prune Leaf Casting Mottle

Prune leaf casting mottle, also called prune mottle and prune leaf mottle, occurs in most prune-producing areas of the western United States and is common on Italian prune in the eastern United States. The disorder also has been observed in other prune cultivars, including Albion, Tragedy, Grand Duke, and Hungarian, but not on Stanley prune. This disorder has not been reported elsewhere.

In the East, chlorotic spots appear on Italian prune leaves in midsummer; the spots turn brownish and become translucent, with indistinct margins. Later, the leaf lamina becomes yellow, and the spots are encircled by a green halo. The spots then become necrotic, and severely affected leaves abscise.

In the West, a chlorotic mosaic pattern is formed. Gray or purple-brown necrotic centers develop in the chlorotic areas. Affected leaves usually abscise, but those that persist turn gray and develop chlorotic leaf margins. On symptomatic trees, fruit set is reduced, and fruit drop prematurely. Disease severity depends on temperature.

The condition is perpetuated only in buds transferred to healthy trees. Thus, it is likely a genetic disorder.

Trees should be propagated from scion sources free of the disorder. Source trees should be examined closely in mid-season and after periods of high temperatures.

Selected References

Helton, A. W. 1962. Interrelationships of bud-perpetuated conditions in prune trees, and variations in these conditions with time. Plant Dis. Rep. 46:523-527.

Parker, K. G., and Gilmer, R. M. 1976. Prune leaf casting mottle. Pages 299-301 in: Virus Diseases and Noninfectious Disorders of Stone Fruits in North America. U.S. Dep. Agric. Agric. Handb. 437.

(Prepared by H. E. Waterworth)

Sweet Cherry Crinkle Leaf and Deep Suture Disorders

Sweet cherry crinkle leaf was described first in California in 1930 as "unproductive cherry." In 1941, Reeves published an account of sweet cherry deep suture in Washington State. Similar symptoms were reported from several European countries and New Zealand. Both disorders cause economic losses—crinkle leaf by reducing fruit yields, and deep suture because packers reject fruit if incidence is high. In North America, crinkle leaf and deep suture probably account for greater annual fruit losses than all virus diseases.

Symptoms

Leaves of trees affected by crinkle leaf are misshapen, with irregularly indented, coarsely serrated margins. In addition, the leaves develop chlorotic or light green patches along the leaf margins (Plate 146). Affected trees produce defective blossoms, and the fruit are small, misshapen, and pointed, with a raised suture (Plate 147). The fruit pedicels vary in length, and yield losses of 50% or more can occur.

With deep suture, there are two leaf symptoms: one form produces long, strap-shaped leaves with a thick-ribbed appearance and irregular margins; in the milder form, leaves are nearly normal in shape, with a darker upper surface and pebbly texture. The name deep suture refers to fruit symptoms, whereby the crease (suture line) is indented (Plate 148). High incidence of deep suture lowers fruit quality considerably.

Several sweet cherry cultivars are known to develop symptoms, including the major cultivars Bing and Black Tartarian and the minor cultivars Black Republican and Chinook. Lambert and Royal Ann (Napoleon) do not develop crinkle leaf or deep suture.

Cause

Both crinkle leaf and deep suture are considered to be genetic in nature. The disorders are bud-propagated but not graft-transmissible to healthy cherry trees.

Control

Propagation material should be taken from trees with a known history of producing quality budwood and no symptoms of these disorders.

Selected References

Blodgett, E. C., and Nyland, G. 1976. Sweet cherry crinkle leaf, deep suture, and variegation. Pages 306-313 in: Virus Diseases and Noninfectious Disorders of Stone Fruits in North America. U.S. Dep. Agric. Agric. Handb. 437.

Kinman, C. F. 1930. A study of some unproductive cherry trees in California. J. Agric. Res. 41:327-335.

Reeves, E. L. 1941. Mottle leaf, a virus disease of cherries. J. Agric. Res. 62:555-572.

(Prepared by M. D. Aichele)

Russet Scab

Russetting, or russet scab, of prune (*Prunus domestica* L.) fruit was first observed in the early 1930s in California. Since then, it has caused severe losses in years with heavy rains during the blossom period. Russet scab is important in the major prune production regions of the world but is not a problem on Japanese plum (*P. salicina* Lindl.). Dried fruit with severe russet must be channeled into the manufacture of juice, which results in lower return to the grower. With the increase in worldwide marketing of fresh prunes, russet scab is one of the major defects requiring effective control.

Symptoms

This condition can be confused with mechanical injury caused when developing fruit rubs against leaves or limbs during high winds (Plate 149). Symptoms are first apparent on green fruit during or shortly after shuck fall. Wax on the affected surface of the fruit is lacking, and the area, first shiny, later develops a brown russet with a netted pattern, a condition originally called "lacy scab" (Plate 150).

Cause

The cause of russet scab on prunes could be microbes causing early injury of the fruit surface, mechanical injury, or chilling in conjunction with an extended wetness period during bloom that prevents the formation of wax by the epidermal tissue. Rain and overhead sprinkling during the bloom period exacerbate the disorder. Removal of the shuck during full bloom reduces the incidence of russetting.

Control

Fungicides (chlorothalonil, captan, or captafol) applied at full bloom control the disease. Sterol biosynthesis inhibitor, dicarboximide, or benzimidazole fungicides are ineffective, and fixed copper sprays give variable results.

Selected References

Corbin, J. B, Lider, J. V, and Roberts, K. O. 1968. Controlling prune russet scab. Calif. Agric. 22(11):6-7.

Michailides, T. 1991. Russetting and russet scab of prune, an environmentally induced fruit disorder: Symptomatology, induction, and control. Plant Dis. 75:1114-1123.

Ogawa, J. M., and English, H. 1991. Diseases of Temperate Zone Tree Fruit and Nut Crops. Pulb. 3345. University of California, Division of Agriculture and Natural Resources, Oakland. 461 pp.

Watanabe, S. 1969. Histological studies on the cause of russet in apples. Bull. Yamagata Univ. 5:823-889.

(Prepared by J. M. Ogawa)

Dead Button

Dead button disorder of Late Montmorency cherry was first reported in 1949 and again in 1956 in two orchards in Utah. Necrosis begins at the stylar end of immature tart cherries and progresses as the fruit matures. The necrotic area varies in size but extends internally to the pit of the fruit. As fruits ripen, the necrotic tissue separates from healthy tissue, leaving a hole in the stylar end. The cause has not been determined but is not graft-transmitted. No spread was observed in the affected orchards, although the same trees were affected year after year.

Selected Reference

Wadley, B. N. 1960. Dead button, a nontransmissible disorder of late Montmorency cherry. (Abstr.) Phytopathology 50:87.

(Prepared by E. I. Zehr)

Nectarine Pox

Nectarine pox disorder is widespread. In California, growers first observed the problem in 1964, and high incidences were observed in 1977, 1978, and 1979 on the cultivars Firebrite, Fantasia, May Red, May Grand, and Royal Giant. In Chile, Australia, and France, the same cultivars are affected. Losses caused by this disorder range from minor reductions in fruit quality to removal of whole orchards.

Symptoms

Symptoms initially are observed about 40 days after anthesis, sometimes as a pale white or light-colored circular spot about 0.5–1.0 mm in diameter. More often, the area is a slightly raised protuberance (0.5 mm), rounded or slightly pointed and 1.0–2.0 mm in diameter at the base. The protuberances, or "warts," are occasionally clustered together, forming a larger irregular, rough, singular outgrowth of the epidermis and subepidermal tissue. At harvest, warts average 3–5 mm in diameter and 0.5–4 mm in height (Plate 151). The raised area is redder than the surrounding tissue which appears normal. Symptoms of nectarine pox are similar to those of the cracking disorder (Fogle and Faust, 1976) and the beady wart disorder (Blodgett, 1976).

Other symptoms described as black scab, scabbing, cracking, pimpling, and red spotting could represent more than one disorder (Plate 152). As first described, dark, scabby patches ranged in diameter from 1.6 to 12.7 mm. Another symptom is raised red bumps that are filled with liquid. On cultivars May Red and May Grand, severely pimpled and toughened skin areas crack as the fruit enlarges rapidly before harvest. On Royal Giant, red pimples and small black, scabby areas form on the same fruit.

Cause

Nectarine pox may result from a nutritional and/or hormonal imbalance. The disorder is severe when the crop load is light, shoot growth and nitrogen, potassium, and/or magnesium levels are excessive, and calcium and/or boron levels are low. There is no evidence that infection by bacteria, fungi, or viruses is involved. Common pesticides (insecticides and fungicides) are not the cause.

Control

Root pruning significantly reduces the incidence of nectarine pox. Incidence is correlated with high nitrogen content in fruit flesh and fruit peel, and root pruning decreases fruit levels of nitrogen, phosphorus, potassium, magnesium, manganese, iron, boron, and zinc but increases calcium.

Based on this information and the fact that this condition is not observed every year in an orchard and is sometimes found only on isolated trees, thinning and irrigation programs that stimulate excessive tree growth and reduce crop production should be avoided. The most susceptible cultivars should not be planted where this disorder occurs.

Selected References

Baugher, T. A., and Miller, S. S. 1991. Nectarine pox: A disorder of nectarine fruit. HortScience 26:310.

Baugher, T. A., and Miller, S. S. 1991. Growth suppression as a control for nectarine pox. HortScience 26:1268-1270.

Blodgett, E. C. 1976. Peach wart. Pages 88-90 in: Virus Diseases and Noninfectious Disorders of Stone Fruits in North America. U.S. Dep. Agric. Agric. Handb. 437.

Fogle, H. W., and Faust, M. 1976. Fruit growth and cracking in nectarines. J. Am. Soc. Hortic. Sci. 101:434-438.

Matsumoto, T., Pool, R., and Shaver, R. 1979. Scab on nectarines. Calif. Dep. Food Agric. Plant Pathol. Rep. 87-88.

Ogawa, J. M., and English, H. 1991. Diseases of Temperate Zone

Tree Fruit and Nut Crops. Publ. 3345. University of California, Division of Agriculture and Natural Resources, Oakland. 461 pp.

(Prepared by J. M. Ogawa and T. A. Baugher)

Peach Skin Discoloration

Peach skin discoloration (PSD), an increasingly important postharvest problem of fresh-market peaches and nectarines, occurs periodically in South Carolina, New Jersey, Virginia, Georgia, and California, as well as in other major peach- and nectarine-growing regions of the world. In different parts of the United States, it is known as ink spot, purple stripe, inking, or brown discoloration. The disorder affects about 12–15% of fresh fruit shipments and causes millions of dollars of loss to the commercial peach and nectarine industry of the United States.

Symptoms

Discoloration can occur in two forms. The typical form first appears as water-soaked areas on the red or blushed surface of the fruit, which eventually turn purplish black (Plate 153). The yellow portion of the fruit surface shows a discoloration described as "grease spot" (Plate 153), such as has been observed in New Jersey in field bins of Cresthaven peaches that had been hydrocooled and placed directly into cold storage. The fruit appears to have gone over too many grading rollers and been extensively bruised. Both forms of discoloration are localized in the epidermis. PSD symptoms are initially expressed after harvest, become more pronounced after cold storage, and remain permanent during the storage life of the fruit.

Peach cultivars that have an intense red stripe rather than an evenly blushed surface are prone to discoloration. The peach cultivars Red Haven, Norman, Garnet Beauty, Sparkle, O'Henry, Elegant Lady, Cassie, June Lady, and Gemfree are highly susceptible. In comparative tests, immature and small fruit were less susceptible than larger, mature fruit, which are softer and more prone to bruising. Bruised areas susceptible to PSD can be detected by removing the fruit from cold storage and holding them at room temperature for 24 hr. Hydrocooling fruit for 20 min in chlorinated water (pH > 8.0) produces blue staining of the red or blushed surface of the fruit. PSD symptoms are temporarily reversed by exposure to mildly acid solutions (0.1 N citric acid), thus implicating the role of anthocyanin pigments.

The brown discoloration induced by field sprays of captan followed by postharvest brushing should not be confused with that of PSD.

Cause

No pathogenic organism has been isolated from PSD-affected fruit, and evidence suggests that abiotic factors stimulate the development of this disorder. Physical injury to the peach or nectarine skin enhances or accelerates the development of PSD on susceptible cultivars. Symptoms on injured skin surfaces are influenced by extremes in pH (acidic or basic), certain metallic ions (Fe^{+2} and Cu^{+2}), and field-applied chemicals. This disorder is most likely caused by the complexing of anthocyanins and tannins in epidermal cells with other cellular components.

Epidemiology

The severity of PSD is a function of epidermal anthocyanin pigmentation as influenced by the environment and mechanical injury. Some cultivars are more susceptible than others, and pre- and postharvest handling operations affect PSD severity. Simulated packinghouse conditions that caused fruit to rub, roll, or physically abrade each other triggered discoloration. PSD has also been induced by chlorinated water (pH > 8.0), excess iron (>10 ppm) in water, leachate from latex rubber drying rollers, ammonia leaks (<1 ppm) from refrigeration systems, and fungicide sprays. The fungicides benomyl, dichloran, captan, and triforine do not individually induce PSD, but postharvest fungicide mixtures in waxes with pH greater than 8.0 may trigger PSD symptoms. Nutritional foliar sprays containing chlorides of iron, aluminum, copper, zinc, and cobalt also have increased the incidence of PSD. Exposure of fruit to high temperatures before postharvest cooling (i.e., leaving the fruit in field bins in the orchard or packinghouse for too long) and the combination of very light rain and high relative humidity within three to five days before harvest also have been reported to stimulate PSD.

Control

Preventive measures for avoiding peach skin discoloration are primarily based on minimizing physical damage to the fruit surface of susceptible cultivars. Placing a foam lid on field bins and lining the bins with polyethylene bubble pads reduces vibration of fruit during transit and prevents injury from direct contact with the wooden sides of field bins. Air shock absorbers are replacing spring suspensions in trucks, thereby reducing vibration during transportation of bins from the field to the packinghouse.

Shifting the pH of the hydrocooler and dump tank containing calcium hypochlorite or chlorinated water from 4.0–6.0 to 6.5–7.0 reduced the concentration of free iron and significantly reduced PSD.

The purplish black discoloration on peaches can be reduced by 0.01 M disodium ethylenediaminetetraacetic acid (EDTA) (pH 3.9) or 0.01 M calcium EDTA (pH 3.5) applied as a postharvest dip. The grease spot form of discoloration is not affected by this treatment.

Selected References

Ceponis, M. J, Cappellini, R. A., Wells, J. M., and Lightner, G. W. 1987. Disorders in plum, peach, and nectarine shipments to the New York market, 1972–1985. Plant Dis. 71:947-952.

Chastagner, G. A., and Ogawa, J. M. 1976. Injury of stone fruits by preharvest captan sprays followed by postharvest treatments. Phytopathology 66:924-927.

Hopfinger, J. A., and Frecon, J. L. 1989. The role of postharvest handling and peach skin discoloration. Acta Hortic. 254:167-173.

Phillips, D. J. 1988. Reduction of transit injury-associated black discoloration of fresh peaches with EDTA treatments. Plant Dis. 72:118-120.

Reeve, R. M. 1958. Histological and histochemical changes in developing and ripening peaches. I. The catechol tannins. Am. J. Bot. 46:210-217.

Ridley, J. D., Van Blaricom, L. O. and Sims, E. T. 1976. Postharvest treatment influences discoloration of fresh peaches. U.S. Dep. Agric. Coop. Ext. Serv., Clemson Univ. Circ. 560.

Van Blaricom, L. O. and Senn, T. L. 1966. Anthocyanin pigments in freestone peaches grown in the Southeast. HortScience 90:541-545.

(Prepared by J. D. Ridley and J. M. Ogawa)

Minor Physiological and Genetic Disorders

Physiological and genetic disorders of temperate zone fruit trees can result in abnormal vegetative growth or the production of unmarketable fruit. These factors are influenced by the genetic components of the specific cultivars, the interaction between rootstock and scion, and climatic, edaphic, and pre- and postharvest cultural conditions.

This section does not cover "June drop," which is a normal abscission of fruits that occurs before endocarp (pit) hardening, preharvest drop, or the alteration of vegetative and fruit growth by use of synthetic or natural growth regulators or pesticides. Nutritional deficiencies and excesses that cause physiological disorders are discussed elsewhere.

False Shot-Hole Disease

Cultivars of sweet cherries and plums, such as Early Burlat cherry and Queen Anne plum, exhibit leaf symptoms similar to those of shot hole caused by *Wilsonomyces carpophilus*. Leaves on these cultivars are riddled with black necrotic spots. Some small spots persist, but others enlarge and eventually abscise. No microorganisms have been isolated from these necrotic regions. This characteristic can be inherited by seedling progeny.

Self-Sterility and Formation of Buttons

Some peach cultivars, such as J. H. Hale and Alomar, are male-sterile, producing little or no viable pollen, a character that is linked to the lack of red pigment in the anther. These cultivars can produce parthenocarpic fruits known as buttons; although the flowers are pollinated, the eggs in the embryo sacs are not fertilized. Pollination without fertilization stimulates the pistils to develop into small, seedless fruits.

Fasciation

This grotesque, abnormal growth habit of shoots is ascribed to a mutation of the terminal meristem induced by natural radiation. Scions irradiated with low dosages of cobalt 60 often give rise to shoots that manifest this abnormal growth. Several terminal shoots on different parts of the same tree may exhibit typical symptoms, which include curved, flat, thickened stems with shortened internodes (Plate 154). Hence, the potential to produce abnormal growth is believed to reside in the vegetative and reproductive cells and may be transmitted sexually or by grafting. Viruses have been isolated from fasciated apple shoots, but whether they are involved with symptom expression is not known.

Graft-Incompatibility

Propagation of some scion cultivars onto a given stock by grafting or budding can lead to poor growth of the scion and eventual death of the tree. Examples are topworking of peaches onto Damson or St. Julien plum stocks and of some cherry cultivars on Stockton Morello cuttings. The affected peach shoot typically displays symptoms of nutritional deficiency because normal transport of nutrients through the graft unions is restricted. In the sweet cherry-Stockton Morello scion-rootstock combination, the shoot may grow well for several seasons and then suddenly collapse or break off at the graft union. The graft failure is attributed to the cells of the rootstock and scion that do not form a good strong union with interlocking fibers.

Excessive overgrowth or undergrowth of the stock relative to the scion is attributed to their different rates of growth and is a symptom of a graft-incompatibility. This condition may eventually lead to death of a tree. However, the constriction at the graft union often dwarfs the tree, which can be considered a desirable trait.

(Prepared by K. Ryugo)

Environmental Disorders

High-Temperature Injuries

Sunburning of Trunks, Scaffold Limbs, and Fruits

Sunburn of trunks and the upper side of limbs exposed to afternoon sun is a common occurrence in stone fruit trees. The temperature of the exposed side can be as much as 4°C higher than that of the lower, shaded side. Likewise, where soil moisture is adequate, the tree trunk is considerably cooler than that of a tree that is water-stressed. The problem can be ameliorated or avoided by 1) leaving the primary and secondary scaffold limbs in a somewhat upright position, 2) pinching back leafy branches on trunks of young trees during summer and retaining short shoots on the upper side of scaffold limbs during dormant pruning so as to provide shade, 3) painting the trunks and upper sides of exposed branches with white latex paint or whitewash, and 4) sprinkling trees with an overhead sprinkler to lower the temperature of the branches by evaporative cooling.

Fruits exposed directly to solar radiation between 2 and 4 p.m. when the air temperature exceeds 40°C will sunburn because the tissue surface temperature can exceed 55°C. The exposed surface of immature apricots and peaches will turn yellow, whereas Kelsey plums develop "heat spots." Evaporative cooling can also be used to reduce fruit temperature.

Leaf Scorch

Leaves exposed to the direct rays of the sun on very hot days become scorched, often beginning at the margins. The direct cause, however, is considered to be increased water stress. High temperatures trigger higher transpiration rates, thereby decreasing leaf water potential.

Pit-Burning of Apricot and Pit-Blackening of Peach

Air temperatures that approach 40°C for a few days when apricots are ripening turn the inner mesocarp tissue brown and gelatinous (Plate 155). Similarly, with many clingstone peach cultivars, the inner mesocarp becomes grayish, especially near the endocarp. Immature apricots exposed to the same temperatures for the same duration do not manifest this symptom. The high temperature tends to accelerate the respiration rate of the ripening fruit to a point that leads to insufficient oxygen in inner tissues. Oxygen from the atmosphere cannot diffuse inward fast enough to meet the need. Pit-burning was induced experimentally in ripe apricots exposed to ambient atmosphere at 40°C, but the symptoms were not reproduced when the atmosphere was oxygen-enriched.

Internal Browning of Prunes

The mesocarp tissue of prunes normally turns from light green to amber and then to brown or even black as fruit ripen. The rate of darkening is hastened by high temperatures and is accentuated if the crop is handled roughly at harvest, because compression of the flesh causes the cells to rupture and the cell contents to become oxidized.

Premature Desiccation of Prunes

The French prune (Petit d'Agen) tends to turn blue and shrivel when exposed to hot, dry winds. The desiccation is attributed to inability of the roots to absorb water as fast as the leaves transpire. Apparently, water previously absorbed by immature fruits is withdrawn as the moisture tension created within the stem increases. The symptom can be reproduced by heating bearing branches with a blowtorch.

Doubling and Spurring of Sweet Cherry and Other Stone Fruits

Exposure of buds to high temperature or direct solar radiation *during floral induction* often results in formation of double pistils. Hence, in the Northern Hemisphere, the problem is more prevalent on the southern and southwestern sides of cherry trees. Evidently, if both pistils of affected sweet cherry flowers are pollinated and the eggs are fertilized, the seeded ovaries fuse along the ventral sutures and develop into doubles (Plate 156). If only one pistil develops, the other forms a spur. Some cultivars are more prone to doubling than others. Apricot and peach buds form double pistils if trees are stressed for water late in the floral induction period. A decreased transpiration rate and high bud temperature have been implicated as causes of this disorder. In cherries, the use of overhead sprinkling when temperature exceeds 30°C has reduced the problem. Another means of amelioration in sweet cherries is to foster lateral budbreak (spur formation) and maintain the shoots in an upright position to maximize shading of flower buds in axils of leaves. Once the pistils begin to differentiate, these treatments are ineffective.

Inadequate Winter Chilling

Flower Bud Abscission, Extended Flowering Period, and Delayed Foliation

After a mild winter, stone fruit trees, especially apricot and peach cultivars with long chilling requirements, drop flower buds just before or at bud swell. Flower buds that set bloom over a long period and extend the harvest period. Lambert sweet cherry trees sometimes bear ripe and green fruits simultaneously following a relatively mild winter. In areas where spring frosts and rains prevail, an extended bloom period of seven to 10 days may be advantageous, because part of the bloom might escape the inclement weather.

Although vegetative buds do not abscise after a mild winter, they do not grow normally. The leaf buds break slowly over a long period and give rise to small, poorly colored leaves, a condition known as *delayed foliation.* The poor vegetative growth reduces yield and increases sunburning of limbs and fruits because of the sparse canopy.

Low-Temperature Injuries

Sudden drops in temperature to below freezing in the fall or early winter before trees become hardy often result in shoot dieback. Injury or death of cells results when water in the vacuole freezes and the ice ruptures the cell membranes. This causes leakage of vacuolar contents and abnormal contact between enzymes and their substrates. When temperature declines gradually, the cells acclimate (become hardy) by transporting water from the vacuole and the cytoplasm to the cell wall and the intercellular spaces. The osmotic concentration of the vacuole increases, and the cytoplasm becomes dense. In extreme cases, the water in the cytoplasm becomes protein-bound. Consequently, ice forms in the cell wall, where crystal formation is relatively harmless, while the protoplast becomes more resistant to freezing. If ice nucleation occurs in the bud, the freezing process spreads basipetally until an entire spur could be injured or killed, because water is contiguous in the cell walls.

In the Northern Hemisphere, especially in higher latitudes, the areas within the tree crotch and on the southwest side of trunks of trees growing on south-facing slopes often crack longitudinally. This cracking is attributed to differential thawing and freezing of water in these zones because the exposed side is warmed by solar heating just before sunset and is cooled rapidly to freezing after sunset.

As flower buds approach anthesis, they become more sensitive to chilling injury. An open flower may survive −2°C for short periods; however, if that temperature persists, the embryo will die and flowers will abscise. When the temperature is just sufficient to cause freezing injury, flowers facing the sky at night are more likely to be injured by heat loss from radiation than are those facing down, which absorb heat from the earth.

The first visible sign of frost damage is browning of the seed coat. Occasionally a light frost will kill an embryo but the fruit may set parthenocarpically. Apple and pear fruit may exhibit a "frost ring" or various degrees of russetting, but these conditions are rarely observed in stone fruits.

Damage from spring frosts can be alleviated by 1) lighting orchard heaters to warm tree tissues through direct radiation while adding heat to the atmosphere; 2) using wind machines to mix the cold air on the orchard floor with warmer air from the inversion layer above the orchard; 3) disking under the cover crop and tamping the soil, because leafy plants intercept the sun's rays and prevent the soil from absorbing heat, while tamping reduces the soil surface from which water can evaporate; and 4) irrigating the orchard with sprinklers, furrows, or basins, because water yields heat as it freezes. Cryoprotectants have been used experimentally to protect tissues from freezing, but results have been unreliable in practice.

Hail Damage

Large hailstones have defoliated trees and even scarred trunks of trees and grapevines. Small hailstones will damage flowers, leaves, and fruits. The scarred tissue does not develop normally and often appears as indentations or streaks covered with callus tissue. Prunes damaged by hail will exude gum, which lowers their quality.

Water-Related Disorders

Cracking of Sweet Cherry

When a water droplet remains on the epidermis for an extended period, it moves through the epidermal cells and into the mesocarp by osmosis. The mesocarp cells increase in volume rapidly, causing the epidermis to stretch. When the epidermal cells reach their limit of elasticity, the epidermis splits (Plate 157). Cracking often begins at the stylar scar, because it is not cutinized and allows for rapid influx of water. All cultivars are susceptible, but to different degrees; Lambert seems to be most resistant to cracking.

Spraying with soluble organic or inorganic calcium salts tends to reduce cracking. Three modes of action have been postulated: 1) calcium ions impede the rate of water uptake by increasing the osmotic concentration; 2) calcium binds with pectins to form an insoluble pectate in the cell wall and middle lamella, which inhibits cell wall expansion; and 3) calcium is incorporated into the cytoplasmic membrane, which decreases its permeability to water. Unless the duration of rain is short, attempts to remove water from the fruit surface with blasts of air from a sprayer or helicopter have not been effective.

Even short exposures of cherries to rain cause microscopic fissures to form in the cuticle. These fissures promote water loss after harvest, so that even though the fruit may appear normal, its quality is usually lowered.

End-Cracking of Prunes

When prune trees are stressed for water early in the season and then irrigated, the distal ends of many fruits crack. Apparently, the epidermis on the distal end of the prune loses its elastic qualities when the fruit is drought-stressed. Consequently, the mesocarp will often separate from the endocarp when the fruit regains its full turgidity from the sudden influx of water. To avoid this situation, irrigate prune trees as needed and do not allow them to become water-stressed.

Side-Cracking of Prunes

Side-cracking of prunes becomes apparent in mid-July on fruits exposed to direct sunlight (Plate 158), especially if the temperature had reached dew point earlier. Exposure to direct sunlight causes the epidermal cells to lose their elasticity. In the early morning hours, the mesocarp cells become fully turgid. If free water from condensation (dew) is absorbed through the skin, the epicarp and mesocarp crack as the cells increase in volume. Cracking of ripe apricot fruit has been observed after a rain.

When immature prunes crack, the mesocarp cells become meristematic again and form an unsightly callus along the wound. When ripe fruits crack, gums and soluble carbohydrates that ooze from the wound serve as a substrate for mold growth. These molds give an off-flavor to the dried product and can produce aflatoxins, some of which are carcinogenic.

Unfavorable Edaphic Factors

Excessive Soil Moisture

Excess soil moisture for an extended period in late fall and/or early spring is often fatal to most fruit tree species. Among stone fruits, the plum rootstock is the most tolerant to "wet feet;" peach, almond, and apricot grafted onto Okinawa or Nemaguard rootstocks seem to be most susceptible.

If roots are exposed to water-saturated soils for a long period, oxygen becomes limiting. Anaerobic respiration leads to formation of alcohol. Anaerobiosis also leads to the formation of hydrocyanic acid (HCN) from the breakdown of amygdalin, a cyanoglucoside. Amygdalin, which is widespread in the genus *Prunus*, is a natural metabolite and is therefore harmless, but as concentrations of hydrocyanic acid and alcohol increase, they cause cells to die.

Insufficient Water

If soil moisture is insufficient during the period of flower initiation (July to October), double pistils are formed in cherries, peaches, and apricots. However, the effect may be indirect, because water stress may lead to elevated tissue temperatures. Lack of water in the spring induces bud drop and may be related to an elevated temperature as the transpiration rate is reduced. Water stress leads to premature defoliation, which may induce flowering in the fall and makes trees susceptible to winter freezing injury.

Studies in Australia showed that regulated-deficit irrigation during the pit-hardening period followed by excess irrigation enhanced fruit growth, but the results were not confirmed in California. Water stress of stone fruit trees at any time during the growing season causes a loss of fruit-sizing potential, because their photosynthetic rate is lower than that of trees receiving adequate water. In extreme cases, fruits on water-stressed trees abscise prematurely.

Growth-Related Disorders

The growth rate of stone fruit is enhanced when a light crop is caused by a late spring frost, chemical fruit thinner misapplication, or alternate bearing, or if two embryos develop within an endocarp. The following physiological disorders often accompany such rapid growth rates.

Splitting of the Endocarp (Pit or Stone)

Large fruits on lightly cropped peach and nectarine trees are apt to possess cracked or split endocarps (Plate 159). This condition is also prevalent in cultivars that ripen early in the season and in those with an elongated rather than a spherical endocarp. Some plum and cherry cultivars may manifest split pits in some years. In nectarines, peaches, and plums, the pit usually splits along the dorsal-ventral suture, but if the crop is light, it may crack in different directions, especially at the apex.

Plum, peach, and nectarine endocarps split because the endocarp tissue ceases to enlarge as its cells begin to lignify. The pit loses elasticity or plasticity, while the mesocarp cells continue to enlarge. Because the inner mesocarp cells are strongly attached to the endocarp at this stage of fruit development, the endocarp is literally pulled apart by pressure exerted by the mesocarp cells. Cultural operations such as excessive thinning of the crop, untimely application of irrigation water, or girdling of the trunk just before or during the pit-hardening period favor endocarp cracking, because these treatments promote a sudden influx of photosynthates and/or water, which increases osmotic concentration and/or turgor of the mesocarp cells.

When conditions for pollination are favorable, so that two ovules develop within a peach or nectarine endocarp, the endocarp will be forced apart while it is still translucent.

Cherries such as the cultivar Rainier have thin endocarps that split when they are bitten or come into contact with the pitting machine. Split pits are a serious problem for the fruit-processing industry because pit fragments in canned or dried products are unsightly and endanger consumers.

Gumming

Gums and mucilages are complex carbohydrates produced by leaves, limbs, and fruits of various stone fruit species. These compounds are normally found in the vascular bundles, but under certain stresses, they become hydrated and cause the bundles to rupture. When this happens, the gums accumulate in pockets or are exuded through weak places in the tissue.

When the endocarp of a peach fruit splits, the vascular bundles rupture in the mesocarp, allowing their gummy contents to ooze to the surface, usually along the ventral suture or from the distal end of the fruit. Fruits that exhibit gumming along the ventral suture usually abscise with the "June drop" fruits, whereas those that exude gums distally at a later stage of fruit development often persist until harvest. Gumming is also caused by insect feeding, bird pecks, and hail damage.

Blossom-End Rot

Desiccation of the distal ends of canning peaches is often associated with large fruits in light crop years. All canning peach cultivars seem to be susceptible, but some more so than others. In extreme cases, the deterioration spreads over the entire surface of the fruit, giving it an unappetizing appearance.

Maturity-Related Disorders

Harvesting fruits prematurely is a common practice among fruit growers who are anxious to ship their crops to obtain the best prices. In contrast, when the market is saturated and prices are low, growers may allow the crop to become overripe. In both cases, the fruits fail to achieve optimum keeping quality and flavor.

Shriveling of Prematurely Harvested Fruits

Fruits harvested prematurely have low contents of soluble solids, causing the cells to have low water potentials. With a low osmotic concentration in the vacuole, fruits tend to lose water readily through cuticular and stomatal transpiration during storage. This water loss causes the fruit to shrivel and renders it inedible.

Cracking of the Fruit Epidermis

Cracking of the epidermis of ripe fruit that has reached the wholesale and retail markets has been attributed to residual free water on the fruit surface as a result of hydrocooling to lower the fruit temperature. The fruit absorbs the free water by osmosis. The epidermal cells cannot stretch to accommodate the sudden uptake of water, so the epidermis splits.

Internal Browning of Stone Fruits

Peaches, nectarines, and plums often turn dark brown around the pit after a long period in ordinary cold storage (Plate 160). The condition is attributed to advanced senescence of cells. The cells apparently lose membrane integrity as they senesce, so that phenolic substrates and enzymes such as polyphenoloxidase react, causing the phenolic compounds to be oxidized and turn brown.

Excess Crowding and Shading

Poor Pigmentation of Fruits

Poor pigmentation following lack of exposure to sunlight may not be a disorder, but it does lower acceptance by consumers. Poor coloration also is associated with lower soluble solids contents. Not all fruit cultivars need direct exposure to sunlight to produce anthocyanins. Completely shading red-fleshed Bing cherry lowered anthocyanin content by only 15%, whereas given the same treatment, the epidermis of the white-fleshed cultivars Royal Ann, Rainier, and Emperor Francis, which normally form a reddish blush when exposed to the sun, remained pure yellow. Although these fruits grown in the dark were larger than exposed ones, their flavor made them inedible.

Poor Flower Bud Formation and Fruit Set

Buds in shaded zones of the tree canopy usually do not form flower buds. Furthermore, flowers that do form in the shady sites are weak and often do not set.

Atmospheric Pollution

Damage to foliage in orchards located downwind from aluminum smelting plants has been attributed to atmospheric pollutants, especially fluorides. Automobile emissions have induced abnormalities experimentally, but orchards along well-traveled freeways have not shown such damage under natural conditions.

Nonbearing Syndrome of Plums

The French prune tree occasionally does not bear a crop, and the leaves turn yellow and abscise. This premature senescence is known as the "nonbearing syndrome." It occurs rarely, and the same tree does not always exhibit the symptoms in subsequent seasons. Because the disorder is unpredictable, little work has been done on it. Early Italian prune trees with a light crop exhibit gum spots and leaf curl, and their fruits do not ripen normally. Whether this condition is related to the nonbearing syndrome of French prunes is not known.

Selected References

Chandler, W. H. 1965. The Deciduous Orchards, 3rd ed. Lea-Febiger, Philadelphia. 492 pp.

Gourley, J. H., and Howlett, F. S. 1947. Modern Fruit Production. The Macmillan Co., NewYork. 579 pp.

Kobayashi, K. D. 1987. Mechanisms of rest and dormancy. HortScience 22:816.

Li, P., and Christersson, L. 1992. Advances in Plant Cold Hardiness. CRC Press, Boca Raton, FL. 332 pp.

Ryugo, K. 1988. Fruit Culture, Its Science and Art. John Wiley & Sons, New York. 387 pp.

Westwood, M. N. 1978. Temperate Zone Pomology. W. H. Freeman & Co., San Francisco. 428 pp.

(Prepared by K. Ryugo)

Nutritional Disorders

Mineral deficiencies and excesses can result in leaf and fruit symptoms that may be confused with symptoms induced by biotic agents. Thus, one must become familiar with symptoms, especially in leaves, caused by nutrient disorders.

Causes of nutrient problems can be many: high or low soil pH that affects the availability of certain elements to the plant; concentration of an element in the soil; competitive uptake of cations; and environmental conditions that limit root growth, such as poor drainage and aeration, unfavorable temperature or soil moisture conditions, and soil texture.

There are three approaches to diagnosing nutrient problems: 1) soil analysis, 2) leaf analysis, and 3) recognition of symptoms. The last approach is the cheapest and quickest method of diagnosis, but it requires experience to become familiar with the various symptoms (Table 5). Soil analysis is useful for predicting nutrient problems that might be encountered after planting the orchard or, if toxicity symptoms appear, to locate the source of the toxic element in the soil. Leaf analysis is probably the most useful approach in diagnosing nutrient problems. The tree integrates the physical, chemical, and environmental factors that affect nutrient uptake. The level of nutrients found in leaves can be compared with levels known

TABLE 5. Classification of Deficiency Symptoms

Plant Part Affected	N	P	K	Mg	Mn	Zn	Fe	Cu	B
Leaves	*Pale yellow*[a]	...	Pale (like N deficiency) "buckskin" color	Marginal chlorosis, *V pattern*	*Interveinal chlorosis*, herring-bone pattern	Interveinal chlorosis, mottled "little leaf"	Overall yellowing, *fine network of green veins*	...	...
	Reddish stems and petioles	Small leaves, poorer growth	Midshoot leaves affected (peach), marginal chlorosis, rolling, graying, bronzing, scorch	Terminal leaves normal, basal leaves affected	Terminal leaves usually normal, basal leaves affected	Terminal leaves show *wavy margins*	*Terminal leaves yellow*, basal leaves greener	*Terminal leaves necrotic, "withertip"*	...
Defoliation	...	...	...	Basal	...	Basal	*Terminal*	*Terminal*	...
Miscellaneous	Affects flower differentiation, fruit set, fruit size, maturity	Rare	Adversely affects *fruit size*	Usually occurs when K is high or soil is acid, especially in young, vigorous trees	In high-pH soils	Delayed bloom and foliation	Lime-induced chlorosis, high pH	*Shoot dieback*, rough bark	*Shoot dieback*, blossom blasting, *fruit abnormalities*

[a] Italicized words designate important symptoms.

TABLE 6. Critical Nutrient Levels in the Leaves of Stone Fruit Trees (Early Summer Samples)[a]

	N[b]		K[c]		Ca	Mg	Na	Cl[d]	B			Zn
Crop	Deficient Below (%)	Adequate (%)	Deficient (%)	Adequate Over (%)	Adequate Over (%)	Adequate Over (%)	Excess Over (%)	Excess Over (%)	Deficient Below (ppm)	Adequate (ppm)	Excess Over (ppm)	Adequate Over (ppm)
Apricots (shipping)	1.8	2.0–2.5	2.0	2.5	2.0	...	0.1	0.2	15	20–70	90	16
Apricots (canning)	2.0	2.5–3.0	2.0	2.5	2.0	...	0.1	0.2	15	20–70	90	16
Cherries (sweet)	...	2.0–3.0	0.9	...	...	...	...	...	20	...	...	14
Nectarines and peaches (freestone)	2.3	2.4–3.3	1.0	1.2	1.0	0.25	0.2	0.3	18	20–80	100	20
Peaches (cling)	2.4	2.6–3.5	1.0	1.2	1.0	0.25	0.2	0.3	18	20–80	100	20
Plums (Japanese)	...	2.3–2.8	1.0	1.1	1.0	0.25	0.2	0.3	25	30–60	80	18
Prunes (processing)	2.2	2.3–2.8	1.0	1.3	1.0	0.25	0.2	0.3	25	30–80	100	18

[a] Adequate levels for all fruit crops: phosphorus (P), 0.1–0.3%; copper (Cu), over 4 ppm; manganese (Mn), over 20 ppm. Leaves are from nonfruiting spurs on spur-bearing trees, fully expanded basal to midshoot leaves on peaches.

[b] Percentage N in August and September samples can be 0.2–0.3% lower than in July samples and still be equivalent. Nitrogen levels higher than underlined values will adversely affect fruit quality and tree growth. Maximum N for Blenheims should be 3.0% and for Tiltons, 3.5%.

[c] Potassium levels between deficient and adequate are considered "low" and may cause reduced fruit sizes in some years.

[d] Excess Na or Cl causes reduced growth at the levels shown. Leaf burn may or may not occur when levels are higher. Confirm salinity problems with soil or root samples.

to be needed for healthy trees (Table 6). These standard levels are usually published by universities and local extension services. When particularly difficult nutrient problems or multiple deficiencies develop, more than one approach may be required for correct diagnosis.

Nitrogen

Since soil particles cannot hold a large reserve of nitrogen (N), growers need to apply it yearly. N is an integral component of chlorophyll, and N deficiency results in pale green leaves and reduced shoot growth (Plate 161). Under low-N conditions, the first leaves to emerge are pale green, and in late spring and early summer, these basal leaves on shoots are small and pale and tend to abscise as the season progresses.

Under low-N conditions, anthocyanin production is encouraged, and fruits are more highly colored than those under high N. Usually, a reddish tinge develops on the petioles and stems as well as on the blades of leaves; under severe N deficiencies, reddish spots can appear in these areas. These reddish symptoms can be pronounced in peaches. In other stone fruits, the main symptoms of low N are pale green leaves, early autumn coloration, and early leaf fall.

N deficiency can be corrected easily by application of N fertilizers. The form of N (ammoniacal, nitrate, urea, etc.) may not be important, because nearly all N fertilizers eventually become nitrate, which is the main form taken up by the tree. Application of excess amounts of N can lead to delay of fruit maturity, susceptibility to certain diseases, and pollution of groundwater. Therefore, apply only amounts that the tree can absorb and utilize, and do not overfertilize, which adds to water pollution problems.

Potassium

Potassium (K) deficiency initially resembles N deficiency in that leaves are slightly pale. In early summer, midshoot leaves of peach trees show curling and rolling of the edges. Leaf margins become chlorotic, eventually developing a marginal scorch. In prunes, the entire leaf becomes chlorotic, and scorching of margins or the entire blade may occur. These symptoms usually are seen on the tops of the trees and, in severe cases in prunes, lead to dieback of the shoots—thus, the term "prune dieback" (Plate 162). In most stone fruits, the underside of the margins of affected leaves develops a gray or bronze cast.

Fruits require considerable amounts of K, so that K deficiency invariably leads to smaller fruit. In prunes, severe K deficiency can also lead to many sunburned fruits, which drop prematurely.

K deficiency is usually corrected by K fertilizers applied in bands on the surface or drilled into the soil. An especially efficient method is to apply K in drip irrigation water or to place K fertilizer under each emitter. Prune growers use sprays of KNO_3 to alleviate some deficiencies temporarily, but the use of K sprays on other stone fruits needs additional research. Topsoil of orchards is usually highest in K, and any cultural practice that allows roots to thrive in the topsoil can help in alleviating K deficiency.

Magnesium

Magnesium (Mg) deficiency is variable but is frequently seen on young, vigorous trees. Mg deficiency is often the result of high K in the tree rather than low Mg in the soil. In vigorous trees, Mg deficiency does not seem to adversely affect yield, fruit size, or leaf size. The characteristic symptom of Mg deficiency is a marginal chlorosis of leaves leaving an inverted green V around the base of the midrib (Plate 163). As the season progresses, the chlorotic area expands and becomes necrotic. In severe cases, the whole leaf becomes affected and the browned leaf abscises.

Mg deficiency symptoms first appear on basal leaves and progressively affect the distal leaves. Defoliation also begins from the base of shoots and progressively affects the leaves above. Young leaves are very seldom affected.

Zinc

Zinc (Zn) deficiency is one of the most common micronutrient deficiencies encountered in stone fruits. It can be found in nearly all growing areas but is most common in coarse-textured and sandy soils.

Zn deficiency is characterized by delayed opening of buds (flower and leaf) in the spring. The onset of bloom and foliation can be delayed a few days to a few weeks and, in very severe cases, even a month. The slowly developing vegetative buds produce small, chlorotic leaves, shortened internodes, and tufts of small leaves along the shoot ("little leaf") (Plate 164).

The affected developing leaves become chlorotic with interveinal mottling. In some stone fruits, especially peach, the young leaves show crinkled or wavy margins (Plate 165). In other species, this symptom may not be obvious. Sometimes in a young, vigorous replant, affected older leaves may abscise. Leaves developing in late spring or early summer may appear normal and mask the initial deficiency symptoms that occurred earlier. Thus, the best time to look for Zn deficiency symptoms is in early spring.

In bearing trees, Zn deficiency not only reduces growth but also can cause small, poorly colored, and sometimes misshapen fruits. In cherries, Zn deficiency could be mistaken for X-disease (caused by an MLO) or vice versa.

Fortunately, Zn deficiency in stone fruits can be corrected by a number of foliar and dormant sprays. Soil treatments may work well, but the responses are erratic.

Iron

Iron (Fe) deficiency can result when there is 1) insufficient Fe in the soil, 2) sufficient but unavailable Fe, or 3) sufficient and available Fe that is not properly utilized in the plant. Most Fe deficiency problems fall into the last category, where, under high soil pH (above 7.0), high bicarbonate (HCO_3^-) occurs in the soil and inhibits the utilization of Fe in the plant. Fe deficiency symptoms that develop under this condition are commonly called lime-induced chlorosis. This is the most common form of Fe deficiency found in orchards.

One symptom of Fe deficiency is a loss of chlorophyll that leads to leaf chlorosis in which the interveinal areas become yellow but the veins remain green, leaving a network of small, green veins (Plate 166). Symptoms develop first on young leaves. Later, the green veins also become chlorotic, and the whole leaf finally turns ivory or even white. Brown, necrotic spots may occur near leaf margins. With time, the lower leaves become chlorotic and develop symptoms. The terminal leaves that first developed symptoms in time abscise, leaving the terminal half of shoots bare.

The loss of chlorophyll also affects fruit. Fruit skins become pale and do not develop the normal blush; fruit size also can be reduced.

Lack of good soil aeration from poor drainage or overirrigation also can contribute to development of Fe deficiency symptoms, and low soil temperatures add to the problem. Corrective measures should aim to prevent waterlogging and overirrigation. Soil pH can be lowered by applying acidifying fertilizers (NH_4NO_3) or adding soil sulfur. Fe chelates as soil treatments have been effective, but the correction is short-lived. Foliar sprays of Fe chelates and other Fe materials have given only partial correction.

Manganese

Manganese (Mn) deficiency symptoms appear when soil pH is about 7.0 and higher. The main characteristic symptom is development of leaf blade chlorosis between the main lateral veins extending from the midrib to the margin of the leaf. The veins remain green, producing a "herringbone" pattern (Plate

167). Tree growth, leaf size, and fruit yield are not significantly reduced unless the deficiency is severe; that is, if 50% or more of the leaves show symptoms.

Mn deficiency can be easily corrected. Soil applications of sulfur or early spring foliar sprays of $MnSO_4$ or Mn chelate should give adequate correction.

Copper

Copper (Cu) deficiency is rare on peaches and nectarines but is seen occasionally on prunes and plums. Symptoms, however, may be confused with those of boron deficiency or even boron excess. Leaf analysis is useful for separating the two deficiencies.

In Cu deficiency, new shoots in the spring appear normal. In early to midsummer, shoot growth ceases, terminal leaves and those immediately below become necrotic, and terminal buds and the stems die (Plate 168). The length of shoot dieback by the end of summer depends on the severity of the deficiency. The bark of the trunk becomes rough in surface texture, and this change is sometimes accompanied by gumming.

Corrective measures have not been fully developed, but early foliar sprays with dilute $CuSO_4$ or Cu chelates are effective.

Boron

Boron (B) deficiency of stone fruits is not very common. It is characterized by death of terminal buds, which results in shoot dieback, and development of many side branches. Since Cu deficiency also causes terminal bud and shoot dieback, it is wise to rely on leaf analysis to separate these two deficiencies. Terminal dieback occurs early in the spring in B-deficient trees and in summer in Cu-deficient trees.

Symptoms of B deficiency can differ somewhat among species. In prunes, considerable witches'-broom growth is evident, and the number of flower buds is very much reduced. Some plums and apricots have necrotic tissue penetrating through the flesh to the endocarp.

B deficiency has been corrected with soil application of borax (200–400 g per tree). Foliar application of a soluble B material has also been satisfactory.

Selected References

Beutel, J., Uriu, K., and Lilleland, O. 1983. Leaf analysis for California deciduous fruits. Pages 15-17 in: Soil and Plant Tissue Testing in California. Div. Agric. Sci., Univ. Calif. Bull. 1879.

Childers, N. F. 1966. Nutrition of Fruit Crops. Hortic. Publ. Rutgers-The State University, New Brunswick, NJ. 888 pp.

LaRue, J. H., and Johnson, R. S. 1989. Peaches, Plums, and Nectarines: Growing and Handling for Fresh Market. Publ. 3331. University of California, Division of Agriculture and Natural Resources, Oakland. 246 pp.

Plucknett, D. L., and Sprague, H. B. 1989. Detecting Mineral Nutrient Deficiencies in Tropical and Temperate Crops. Westview Trop. Agric. Ser., No. 7. 553 pp.

(Prepared by K. Uriu)

Index